MATERIALS
PROCESSING
BY CLUSTER ION BEAMS

MATERIALS PROCESSING
BY CLUSTER ION BEAMS

History, Technology, and Applications

Isao Yamada

CRC Press
Taylor & Francis Group
Boca Raton   London   New York

CRC Press is an imprint of the
Taylor & Francis Group, an **informa** business

CRC Press
Taylor & Francis Group
6000 Broken Sound Parkway NW, Suite 300
Boca Raton, FL 33487-2742

First issued in paperback 2019

ISBN-13: 978-1-4987-1175-3 (hbk)
ISBN-13: 978-0-367-87229-8 (pbk)

Visit the Taylor & Francis Web site at
http://www.taylorandfrancis.com

and the CRC Press Web site at
http://www.crcpress.com

*To my wife, Eiko, and my longtime friend, Allen Kirkpatrick.
Together they made possible my research
efforts and publication of this book.*

Contents

Acknowledgments

The author expresses his thanks to the many individuals who have been associated with his 40 years of research and development at Kyoto University and University of Hyogo (formerly Himeji Institute of Technology). In particular, contributions by Jiro Matsuo (Kyoto University), Noriaki Toyoda (University of Hyogo), Takaaki Aoki (Kyoto University), and Toshio Seki (Kyoto University), who worked with him for many years, have been enormously important in making cluster ion beam technology successful.

He also thanks Allen Kirkpatrick, founder of Epion Corporation, for his support of the fundamental research activities from the challenging earliest stages of cluster ion beam technology development in Japan and for his long collaboration in the creation of commercial gas cluster ion beam (GCIB) equipment in the United States. He is also grateful to Allen Kirkpatrick for careful reading of the chapters of this book and for providing much informed criticism and offering many suggestions that have been incorporated into the final publication.

The author thanks the many people who joined the Ion Beam Engineering Experimental Laboratory of Kyoto University for cluster beam research as visiting professors and who contributed to making it possible to bring the technology from fundamental research to industrial applications. They include Dr. Jürgen Gspann of the Institut für Kernverfahrenstechnik der Universität und des Kemforschungszentrums Karlsruhe; Dr. Zinetulla A. Insepov of the Levedev Physics Institute; Dr. Erin C. Jones of the University of California, Berkeley; Dr. Rafael R. Manory of Ben Gurion University; Dr. Claus E. Ascheron of Leipzig University; Professor Max L. Swanson of the University of North Carolina; Dr. Seok-Keun Koh of Rutgers University; Professor Jan A. Northby of the University of Rhode Island; Dr. Michael I. Current of Silicon Genesis; Professor Leonard L. Levenson of the University of Colorado; Professor Rolf E. Hummel of the University of Florida; Professor Marek Sosnowski of the New Jersey

Institute of Technology; and Professor Ronald P. Howson of
the Loughborough University of Technology.

Valuable advice and guidance were also given by
Dr. Otto F. Hagena of Kemforschungszentrums Karlsruhe,
Professor Gilbert D. Stein of Northwestern University, and
Dr. Walter L. Brown of Bell Laboratories.

Thanks are also due to collaborators at Fujitsu, Ltd. and
Nissin Ion Equipment Co., Ltd., who participated in the pioneering development of decaborane implantation under the
Risk Taking Fund for Technology Development Program of
the Japan Science and Technology Agency. Throughout the
history of cluster ion beam work, development has been generously supported by many Japanese government organizations: Ministry of Education, Japan Science and Technology
Agency, and Japan Society for Promotion of Science. Special
thanks are due to the New Energy and Industrial Technology
Development Organization (NEDO), which for a total of nine
years supported two large national projects for cluster beam
technology development: Cluster Ion Beam Process Technology
and Advanced Quantum Beam Process Technology. The success of these projects was greatly aided by the Osaka Science
and Technology Center in Osaka, Japan, which handled the
complicated administrative coordination between government and project members. Contributions by many partici-
pants, including papers by the project members, are listed at
the website http://clusterion.jp/. Last, the author thanks Luna
Han of Taylor & Francis for her patient support in making
this book possible.

Preface

The book is intended to be used as a text in university-level materials processing courses or to complement the texts traditionally used in courses on general ion beam technology, including those for ion beam beginners. It is also strongly recommended for researchers in fields involving ion and plasma technologies or for those who want to go into new areas of surface processing. The book is useful for learning contemporary physics, materials science, surface engineering issues, and the nanotechnology capabilities of cluster beam processing, which is becoming widely employed within high-technology industries.

Cluster beams can be produced from solid materials or gases. Solid source clusters are of much interest for materials science investigations because they can represent a bridge for studying differences between individual atoms and bulk materials. However, this book emphasizes clusters created from gas sources, which have emerged as versatile and powerful tools for surface processing, particularly at atomic levels, and as useful for diverse applications in the fabrication of nanotechnology devices.

The author of this book was the originator of the gas cluster ion beam (GCIB) concept. He has been actively involved with GCIB technology since the beginning, and is uniquely qualified to document the historical background, the progress of technological development, and the current status of the technology. Because of his in-depth knowledge of GCIB and related fields, this book will help both newcomers and practicing professionals within science and engineering communities to fully understand the importance of cluster ion beams and their applications.

The book presents a history of the technology, underscoring how this new area relates to more familiar work. It also explores the challenges involved in expanding the technology to potential uses, in the areas of nanotechnologies such as surface modification and etching, ultrasmooth surface formation, high-quality thin-film formation, and shallow junction formation based upon remarkable characteristics of cluster beam interactions with solid surfaces. It further examines expanding applications in surface analysis tools such as x-ray photoelectron spectroscopy (XPS) and secondary ion mass spectroscopy (SIMS) and discusses surface modifications on biomaterials, which are being implemented to overcome existing technological limits.

In Chapter 1, an overview and the history of ion beam technologies are presented, starting from the earliest discovery of monomer ions and monomer ion beams and continuing on through the introduction of gas cluster ion beams and polyatomic ion beams almost a century later. Views and experiences are described concerning various aspects of the history of ion implantation for the semiconductor industry that are useful to us even now to help us appreciate what was involved in advancing an unknown new concept into a fully mature technological field. Innovative technologies such as plasma immersion ion implantation and gas cluster ion beam processing evolved because of these historical experiences.

In Chapter 2, the history and milestones of the development of cluster beams are reviewed. An understanding of the content of this chapter should be helpful to both researchers seeking to further advance the technology and newcomers to the field because the author provides insight concerning the many contributions that were needed in order for gas cluster technology to become successful. Moving the technology forward from initial feasibility through to transfer to industry and commercialization required not only individual efforts, but also large, carefully coordinated collaborative programs. Individual efforts alone would not have been sufficient. The chapter describes many examples of the conscientious collaborations that brought the efforts to successful outcomes.

Chapter 3 discusses the development of sources for producing cluster beams from solid materials. As has also been the case with GCIB, cluster beams generated from solid materials have become important subjects for fundamental materials research. Methods to produce such beams are relevant topics for future expansion into technological applications. Several examples are introduced in the chapter.

In Chapter 4, the engineering characteristics of gas cluster ion beam equipment are described. The supersonic expansion of substantial volumes of gas through a nozzle into high vacuum, which is a basic GCIB requirement, is almost fundamentally inconsistent with what is normally considered to be good vacuum practice. Suitable approaches to address this issue were developed, and practical configurations for GCIB systems are now well established. Examples of the equipment that has been produced are presented as references, which may be useful as GCIB is used for new applications in the future.

In Chapter 5, cluster ion–solid surface interaction kinetics are described. The fundamental characteristics of the interactions of cluster ion beams with solid surfaces are quite different than those associated with normal monomer ion beams. When GCIB emerged, ions composed of large aggregates of atoms were new to the ion beam community, and they introduced subtle new phenomena for which explanations were not obvious. Examples of such phenomena included low-energy effects, lateral sputtering behavior, and extraordinary chemical reaction effects. Molecular dynamics simulations, in combination with experimental observations and confirmations, have been highly successful in providing the required explanations.

In Chapters 6 to 8, sputtering, implantation, and ion-assisted deposition, which utilize the unusual characteristics of cluster ion beam bombardment on solid surfaces, are reviewed. Also described are the characteristics of GCIB and related surface processing techniques for smoothing, shallow implantation, and preparation of high-quality thin films.

In Chapter 9, representative examples of industrial applications that are emerging for GCIB are introduced. Included among these examples are nanoscale etching and smoothing techniques that can be realized only by employing cluster beams and GCIB infusion doping, which involves a number of ultrashallow effects that are radically different from those associated with conventional ion implantation doping processes. GCIB low-energy effect and lateral sputtering behavior are used to produce SIMS and XPS analytical instrumentation with superior capabilities for analyzing complex structures and high-mass organic materials. Applications of GCIB for the purpose of improving the biological response characteristics of materials used for biomedical implants are becoming a rapidly growing opportunity for the technology. Several examples are presented.

The author anticipates continuing advances in the science and applications of cluster beams because the technology is still young and many exciting opportunities remain to be explored.

Isao Yamada

Author

Isao Yamada is presently a visiting professor at the Graduate School of Engineering of the University of Hyogo in Japan. He is professor emeritus at Kyoto University, where he was director of the Ion Beam Engineering Experimental Laboratory. He earned his PhD degree in electrical engineering from Kyoto University and has held visiting scientist appointments at the FOM Institute in Amsterdam, the Massachusetts Institute of Technology, Cornell University, and Northwestern University.

The principal areas of his research interests over the past 40 years have spanned fundamental physics to practical applications of materials processing by ion beams, with particular emphasis on very low-energy ion–solid interactions. He was the originator of the concept for processing materials by gas cluster ion beams. Since the beginning, he has remained actively involved with development of the technology, which is now becoming one of the basic tools of the emerging field of nanoscale fabrication.

1

Ion Beam Technology
Overview and History

1.1 Overview

Cluster ion beams have become powerful and versatile tools for processing surfaces of materials. Development of cluster ion beams has followed extensive development of conventional ion beam technology, specifically the advances made in equipment and processes for implantation doping of semiconductors. Throughout many decades of the twentieth century, and starting even earlier, great progress was made in advancing ion beam techniques and equipment. The developmental history of ion beams in general and ion implantation in particular represents an excellent example of how the collective contributions of a great many individual researchers and organizations can lead to enormous progress in a new field of technology. Figure 1.1 outlines a simple history of ion implantation development.

Ion implantation technology can be divided into three categories based on the equipment concepts [1]. The first category is conventional ion beams, which can consist of either atomic or molecular ions. In conventional ion implantation, ions produced under vacuum within an appropriate source are extracted by electric fields to form an energetic beam that can then be accelerated or decelerated by additional electric fields. The ions pass

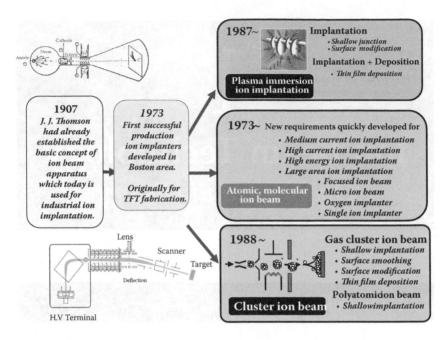

Figure 1.1 Development history of industrial ion equipment.

through a sophisticated beam transport system that typically includes provision for mass analysis and beam scanning; the ions then are implanted into the solid material target.

The second category is plasma immersion ion implantation (PIII). This technology was initially developed to be a versatile tool for metal surface modification and was aimed at avoiding the inherent line-of-sight characteristic nature of traditional ion beam processes [2]. In PIII, targets to be implanted are placed directly within a plasma source and are then pulse biased to a high negative potential. A plasma sheath forms around the target, and ions are accelerated across the whole sheath region so as to bombard the entire target simultaneously. Since its beginning, the technology has been continuously improved and refined for many applications, not only for metals, but also for semiconductors and organic materials.

The third category is that of cluster ion beams [3]. This technology employs ions of clusters consisting of a few hundred to many thousands of atoms. In cluster ion beam bombardment of solid surfaces, the concurrent energetic interactions between many atoms comprising a cluster and many atoms at a target surface result in highly nonlinear sputtering and implantation effects. In addition to their applications for implantation, cluster ion beams have become useful surface-processing techniques in a number of nanotechnology areas that employ their unique very low-energy irradiation effects, lateral sputtering effects, and high chemical reactivity effects.

1.2 History

The history of ion beams began near the end of the nineteenth century. In 1886, Eugen Goldstein observed unknown particles during an experiment with a discharge tube [4]. The discharge tube, as shown in Figure 1.2a, contained an anode (a) and a perforated cathode (K) connected to an induction coil via wire (d). During discharge operation at low gas pressures, a luminous particle stream was observed behind the cathode

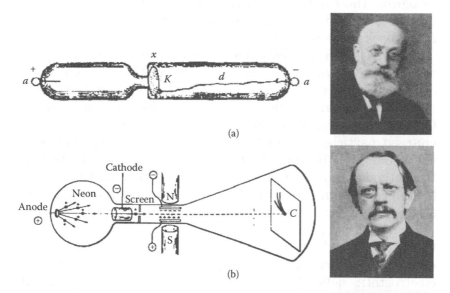

(a)

(b)

Figure 1.2 Early experimental equipment used for discovery and investigation of ion beams. (a) Gas discharge tube by Eugen Goldstein. (b) Joseph J. Thomson's apparatus as depicted by George Gamow.

proceeding in a direction opposite to that of the cathode rays. Behind the perforated cathode, beams of light could be seen streaming through the holes. Goldstein called these unknown rays *Kanalstrahlen*, or "canal rays." Goldstein's canal rays created little notice until 1897, when Wilhelm Wien used powerful magnetic fields to deflect the rays and showed that some of them were positively charged. The observations made by Goldstein and then by Wien can be considered to mark the origin of ion beams [5, 6].

Following these two achievements by Goldstein and Wien, Joseph J. Thomson also conducted experiments on canal rays using an improved apparatus of his own design. Thomson is famous for identifying and characterizing the cathode rays as electrons, but he also studied the canal rays by employing electric and magnetic fields to determine charge-to-mass ratios of the positively charged species from many gases. The main purpose of Thomson's research at the time was to obtain clearer and more detailed images of Goldstein's canal rays. By analyzing trajectories of the canal rays, he was the first to recognize the existence of different isotopes of an elemental gas. Details of his research and contributions are described in many books and articles [7–10].

In 1907, Thomson showed a version of the positive ray apparatus in the *Philosophical Magazine* under the title "On Rays of Positive Electricity" [11, 12]. A simple sketch of Thomson's apparatus, shown in Figure 1.2b [13], was given by George Gamow, who described the operation as follows: "Positively charged particles, originating in the gas discharge between the anode and the cathode, passed through a hole drilled in the cathode, and entered the region of electric and magnetic fields oriented in the same direction." The charged particles struck a fluorescent screen that allowed them to be observed, and Thomson was able to identify the nature of the particles by their actions under the electric and magnetic fields. Thomson's equipment, which employed a gas discharge ion source, an electrostatic deflection system, and a magnet, established the basic concepts of ion beam apparatus that are still in use today. Moreover, his work led to increased interest in the production and manipulation of ion beams throughout the following century. Another important historical contribution at the time by Thomson, together with the help of Francis Aston, was the creation of the first mass spectrograph demonstrated in 1912 [14, 15].

By the 1950s and 1960s, industrial applications of ion beams had progressed extensively and included a wide range of diverse processes. The advancement of ion beam technology has always been strongly related to both equipment development and process application studies. Among the most significant advances toward industrial utilization of ion beam technology were the invention of ion implantation by Russell Ohl, who first described it in U.S. Patent 2750541, "Semiconductor Translating Device," filed January 31, 1950, and William Shockley's U.S. Patent 2787564, "Forming Semiconductive Devices by Ionic Bombardment," filed October 28, 1954.

When Ohl invented ion implantation, the ionic species he studied included air, oxygen, hydrogen, nitrogen, helium, argon, carbon monoxide, and even chloroform. He suggested in his patent that the most advantageous use of his approach was likely to be helium or argon in the production of materials for photocells and high-voltage rectifiers. He thought that when ions entered a semiconductor crystal surface, they changed the electrical characteristics. Three years later, Shockley disclosed the process of ion bombardment using Group III or V ions having energies sufficient to penetrate the surface of a semiconductor to produce a change in conductivity type, as opposed to only a change in degree of conductivity, as suggested in Ohl's patent. Shockley was the first to recognize that the conductivity-type characteristics of semiconductor materials could be changed by the selection of the implant doping ions, and that dopant depth distributions could be altered by varying the energy of the ions. From a historical viewpoint, Shockley's results are considered the first useful application of ion implantation [16], but identical ion implantation apparatus, shown in Figure 1.3, appeared in the patents of both Ohl and Shockley. The equipment was simple, without mass separation or beam scanning, but it presented the fundamental concept of the technology for semiconductor device fabrication.

The early history of ion implantation into semiconductors, including the genealogy of the commercial implanter manufacturing companies and some of the people who contributed

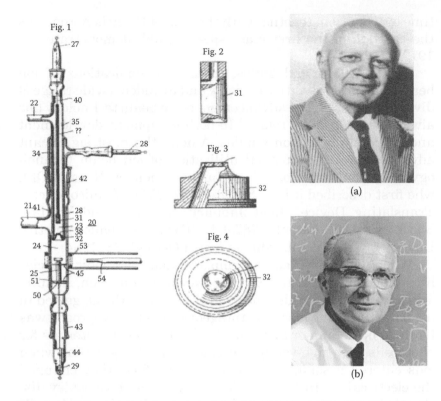

Figure 1.3 Apparatus shown in original ion implantation patents submitted by (a) Russell Ohl in 1950 and (b) William Shockley in 1954. In both patents, the same apparatus on the left side of the figure was presented.

to the field, has been described in the records of the Ion Implantation Technology conference series [17–20]. The problems that were experienced in the ion beam development and the way these problems were handled should be of interest to future generations because similar situations will be encountered again. Lienhard Wegmann stated in his tutorial talk at the 1984 Ion Implantation Technology conference that "in spite of some scepticism as to the usefulness of such a sophisticated, expensive yet brutal method for simply doping delicate semiconductors, in the early sixties, quite a number of laboratories began research in this field" [21]. Charles McKenna described in one of his review papers that "from the beginning, implanters were both competing with and linked to the output from furnaces, so even the initial applications required high throughputs. And the development costs were difficult to recover from a customer" [19]. He emphasized that the costs were related to both equipment and product developments and that these were contributing additional financial pressures to the equipment manufacturers. Even though considerable early progress in equipment development had been made, the development efforts had to be continuously maintained and expanded to meet the ever-more-demanding requirements as the process applications progressed. Even today, one never encounters a shortage of development challenges.

Several published papers have provided insights concerning the early development of practical ion implantation equipment [22–24]. Among such papers are some by Peter Rose, who in 1971 founded Extrion Corporation, the company that developed implantation machines that by 1973 had become the most successful early production equipment for semiconductor doping [23]. Figure 1.4 shows the basic configuration of the early implanter system developed by Rose's team at Extrion. Extrion quickly established a leadership position in the commercial ion implanter business, and more than 500 of its machines were installed between 1970 and 1980 [19]. The success of these

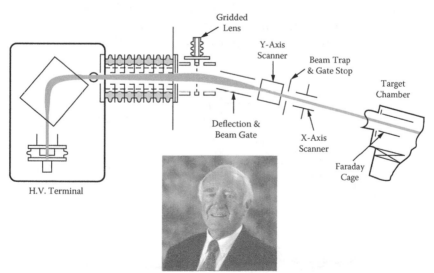

Figure 1.4 Sketch of the configuration of the early production implantation equipment developed under Peter Rose at Extrion. (Courtesy of Peter Rose.)

machines was attributed not only to good system design, but also to the practical wafer-handling capabilities, low contamination characteristics, system reliability, and quality of the service provided by Extrion [21]. While the importance of ion implantation equipment to semiconductor device manufacturing was growing, many other ion beam techniques, such as sputtering and ion-assisted film deposition, were also growing in importance for a wide range of industrial applications.

The second category of ion implantation technology, plasma immersion ion implantation, was initially employed to produce hard, wear-resistant surfaces. Figure 1.5 shows the concept for the original plasma source ion implantation equipment as it was presented in U.S. Patent 4764394, "Method and Apparatus for Plasma Source Ion Implantation," filed by John Conrad on January 20, 1987. In the apparatus as shown, a beam of electrons produced by a source (31) was injected into a cylindrical chamber (12). Magnet bars (32) distributed around the outer periphery of the chamber wall maximized collisions between the electrons and the ambient gas within the chamber. The target to be processed was mounted on a support arm (22) and connected to high-voltage pulse power delivered by power supply (24).

The plasma immersion technique was initially recognized to be capable of enabling simultaneous treatment of large industrial components and complex shapes without requiring

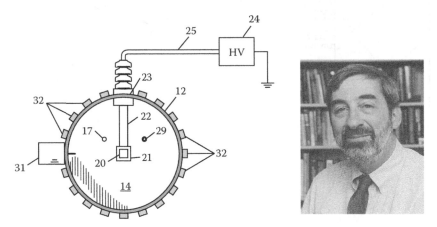

Figure 1.5 Sketch of plasma source ion implantation apparatus shown in the patent by John Conrad at the University of Wisconsin–Madison. (Courtesy of John Conrad.)

component manipulation. As the technology progressed, plasma immersion ion implantation also became known as plasma source ion implantation (PSII) or plasma-based ion implantation (PBII) [25]. At present, the principal semiconductor application of plasma immersion implantation is for doping of poly-Si gate electrodes and contacts in dynamic random access memory. Development for ultrashallow junctions and for use in 3D transistors is in progress [26].

The third category of ion beam equipment technology, that associated with cluster ions, was developed to meet various nanoscale requirements that are normally not satisfied by the first two categories discussed above. Even though conventional atomic and molecular ion beams have long been employed in many implantation, etching, and ion-assisted film deposition applications, recent technological advances have created challenging requirements for new manufacturing methods offering superior atomic-level accuracies in many types of nanotechnology structures with characteristic dimensions of a few nanometers or less. As examples, shallow junction formation, low-surface-damage processing, ultrasmooth surface creation, and atomic-level etching have become important in the fabrication of advanced electronic, magnetic, and optical devices. For such applications, very low-energy ion beam processing is required, but it is difficult to employ such low-energy beams because of beam transport limitations due to space charge effects. Cluster ion beams have provided solutions.

Development of gas cluster ion beam techniques for surface processing has followed upon extensive experimental work and fundamental research studies related to supersonic nozzle beams and nucleation phenomena. And of course cluster ion beam development has also been strongly influenced by the development of other ion beam equipment and processes. Work on cluster ion beam process technology was started around 1988 under the direction of Isao Yamada at the Ion Beam Engineering Experimental Laboratory of Kyoto University after intense gas cluster beam formation from room temperature supersonic nozzles had been observed. Details of the subsequent developmental history are discussed in Chapter 2. The basic configuration of what became known as a gas cluster ion beam (GCIB) device is shown in Figure 1.6.

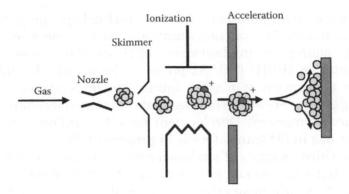

Figure 1.6 Schematic of basic GCIB equipment.

Ion beam currents in early gas cluster equipment were very low, sometimes only on the order of a few nanoamperes, and the factors responsible for beam intensity limitations were not well understood. Further, to generate and transport these very low-cluster ion currents, GCIB equipment typically had to use costly high-capacity vacuum pumps. By 1995, scientists recognized the impracticality of using substantially greater gas flows to increase cluster ion beam currents to the levels required for production processes. No other groups or institutes in the world had yet paid attention to the concept of cluster ion beam equipment or to possible uses of gas cluster ions for surface processing. Several Japanese equipment companies that, at the time, were leading advanced technology developers declined to undertake the task of developing industrial GCIB equipment. Reasons included concerns that the technology might not become sufficiently important, the development times might be unacceptably long, the machines might be too costly and might not be marketable, and the companies did not feel that they had the proper expertise to perform the necessary development starting at such an early stage.

Nevertheless, in collaboration with Yamada's group at Kyoto University and in support of work sponsored by the Japan Science and Technology Agency (JST), Epion Corporation, under the direction of Allen Kirkpatrick, began developing commercial GCIB equipment in the United States [27]. The challenges were considerable, but the Kyoto University and Epion collaboration efforts were successful, and the

Figure 1.7 First GCIB equipment (Epion Model GCIB 30) delivered to Kyoto University under the direction of Allen Kirkpatrick at Epion Corporation. (Courtesy of Allen Kirkpatrick.)

collaborators were able to increase cluster generation, improve efficiency of cluster ionization, and optimize beam transport without increasing gas consumption or pumping requirements. Cluster ion beam currents of several hundred microamperes on target became possible with source gas flows that could be handled by standard vacuum pumps. In 1997, Epion Corporation, with the help of Minoru Nakajima of the import/export company Techscience Ltd., Kosigaya city Saitama, Japan, delivered the first commercially built GCIB R&D system to Kyoto University. Figure 1.7 shows the historically first commercial GCIB equipment, Epion Model GCIB 30, which is still in R&D use at the University of Hyogo.

The first successful commercial production GCIB system, known as the GCIB Ultra-Smoother™, was delivered by Epion Corporation beginning in 1999. One of these units is shown in Figure 1.8. Systems were initially designed to produce extremely smooth surfaces on critical layers used in read-write heads for computer hard disks [28, 29]. They were able to deliver 50 µA of argon cluster ion beam current at energies up to 30 keV, and they provided automated process setup and control, cassette-to-cassette handling of substrates through a vacuum load lock, and mechanical scanning to produce uniform and reproducible processing of substrates up to 200 mm

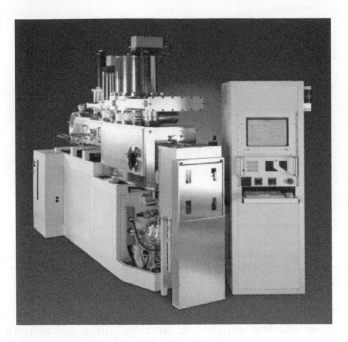

Figure 1.8 Epion GCIB Ultra-Smoother™.

in diameter or square. The cluster beams were magnetically filtered to eliminate monomer ions and very small, multiatom species that can produce detrimental effects. Ultra-Smoother™ were also used for applications in optics and in corrective trimming of surface acoustic wave (SAW) devices and film bulk acoustic resonator (FBAR) devices used in cell phones and GPS systems [30].

As achievable cluster ion beam currents increased, GCIB equipment known as the nFusion™ GCIB Production System, shown in Figure 1.9, was introduced by Epion Corporation in 2005 for automated processing of semiconductor wafers up to 300 mm in diameter [31, 32]. The nFusion GCIB system can be used for surface smoothing, thin surface modification, ultrashallow junction formation, thin-film deposition, etch processing of semiconductors, and processing of metals and insulating materials. Figure 1.10 shows the early evolution of achievable GCIB beam current. Available GCIB beam current levels are now sufficiently high that many production applications are economically practical.

Figure 1.9 Epion 300 mm GCIB nFusion™ system.

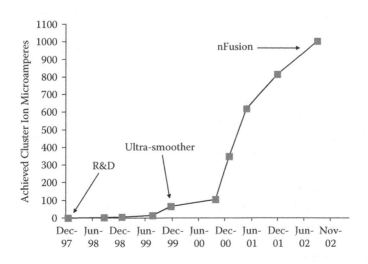

Figure 1.10 Evolution of achievable GCIB current.

To promote industrial applications of GCIB processes, two large-scale projects were conducted under sponsorship from the Japanese government. Beginning in 2000, a 5-year project called Cluster Ion Beam Process Technology, involving development of basic industrial technology utilizing GCIB, was funded under a contract from the Ministry of Economy and Technology for Industry (METI) and was directed by the Ion Beam Engineering Experimental Laboratory at Kyoto University [33]. The project team comprised two groups: one was responsible for developing high-current cluster ion beam equipment; the other was responsible for processing development, including semiconductor surface-processing techniques, methods for smoothing and etching of hard materials, and deposition approaches to the formation of high-quality thin films. In 2002, a second major GCIB project with special emphasis on nanotechnology applications was started under a contract from the New Energy and Industrial Technology Development Organization (NEDO). This project was aimed at developing size-selected cluster ion beam equipment and very high-rate and nondamaging processing for magnetic and compound semiconductor materials. The total budget of the two projects was US$23 million. In support of these projects, GCIB equipment development for the industrial tools was done by Epion Corporation Japan, which was a subsidiary of Epion Corporation U.S. In 2007, a Japanese semiconductor equipment company, Tokyo Electron Limited (TEL), acquired Epion Corporation with the objective of employing GCIB technology for high-volume production applications in semiconductors and emerging nanotechnology markets. Accordingly, the company was renamed TEL Epion, Inc.

Over the 20 years of cluster ion beam investigations, low-energy surface interaction effects, lateral sputtering phenomena, and high-rate chemical reaction effects were explored experimentally and explained by means of molecular dynamics modeling. The fundamental results concerning ion–solid surface interactions have become the basis for a considerable amount of application technology. Cluster ion beams have been applied in nanoscale processing such as shallow junction

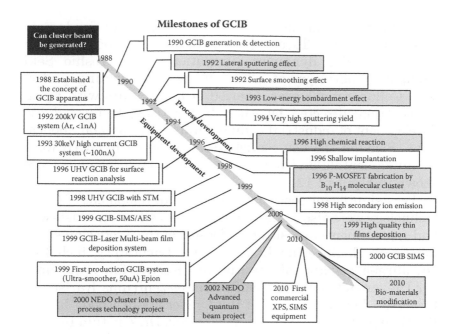

Figure 1.11 Milestones of cluster ion beam technology development.

formation, low-damage surface modification and etching, ultrasmooth surface formation, and high-quality thin-film formation. Since 2007, GCIB development has expanded into analytical instrumentation and also into biomedical device applications. Very low-energy irradiation effects of GCIB have become extremely useful for organic device material analyses by x-ray photoelectron spectroscopy (XPS) and secondary ion mass spectroscopy (SIMS). In the area of biomaterials, GCIB is being employed to modify surfaces of various materials used in biomedical devices with the objective of improving biological responses to the surfaces, for example, to promote faster healing, more rapid bone growth, or stronger integration. Figure 1.11 shows milestones of the first 20 years of GCIB development.

The most important technical fundamentals of GCIB were established in the early work conducted at the Kyoto University Ion Beam Engineering Experimental Laboratory. Many important contributions to the original work at Kyoto

University were made by Jiro Matsuo (process fundamentals) [34], Noriaki Toyoda (lateral sputtering) [35], Takaaki Aoki (molecular dynamics simulations) [36], Toshio Seki (equipment development) [37], and several visiting professors. Collaborations over a period of many years with Epion Corporation, directed by Allen Kirkpatrick, facilitated the development of industrial equipment and commercial applications. Some of the key contributors to the early GCIB work on fundamentals and industrial apparatus development are shown in Figure 1.12.

Very significant contributions to the development of GCIB technology were also made by a series of excellent scientists who had appointments as visiting professors at the Kyoto University Ion Beam Engineering Experimental Laboratory, including Dr. Jürgen Gspann of Institut für Kernverfahrenstechnik der Universität und des Kemforschungszentrums Karlsruhe, Professor Jan A. Northby of the University of Rhode Island, Dr. Michael I. Current of Silicon Genesis, Dr. Zinetula Insepov of the Kazakh State University, Professor Leonard L. Levenson of the University of Colorado, Professor Rolf E. Fummel of the University of Florida, Professor Marek Sosnowski of New Jersey Institute of Technology, and Professor Ronald P. Howson of Loughborough University of Technology. Valuable advice and guidance were also given by Dr. Otto F. Hagena of Kemforschungszentrums Karlsruhe, Professor Gilbert D. Stein of Northwestern University, Dr. Walter L. Brown of Bell Labs, and many others.

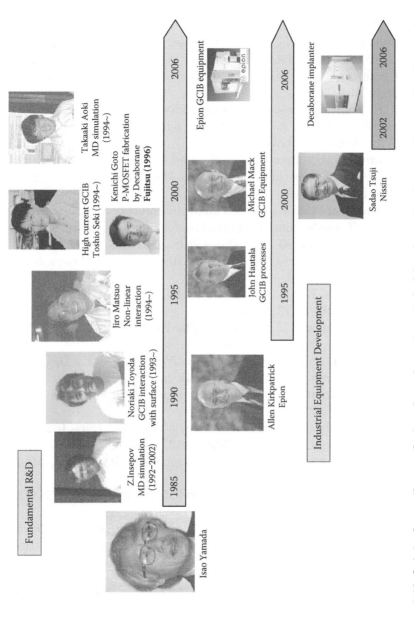

Figure 1.12 Original contributors during the embryonic period of the cluster ion beam technology development.

References

1. I. Yamada. Japan's contribution to ion beam technologies. In *18th International Conference on Ion Implantation Technology IIT 2010*, eds. J. Matsuo, M. Kase, T. Aoki, and T. Seki, 1–8. AIP Conference Proceedings 1321. American Institute of Physics, New York, 2010.

2. J. R. Conrad. Introduction. In *Handbook of Plasma Immersion Ion Implantation and Deposition*, ed. A. Anders, 1–26. John Wiley & Sons, Hoboken, NJ, 2000.

3. I. Yamada, J. Matsuo, N. Toyoda, and A. Kirkpatrick. Materials processing by gas cluster ion beams. *Mater. Sci. Eng.*, R34, 231–295, 2001.

4. E. Goldstein. Ueber eine noch nicht untersuchte Strahlungs form an der Kathode inducirter Entladungen, Sitzungsbericht der Berl. Akad., 25. Juli 1886. *Ann. Phys. Chem.*, 64(1), 38–48, 1898.

5. K. Wien. 100 years of ion beams: Willy Wien's canal rays. *Braz. J. Phys.*, 29(3), 401–414, 1999.

6. M. Hedonus. Eugen Goldstein und die Kathodenstrahlen. *Phys. Blater.*, 56(9), 71–73, 2000.

7. H. A. Boorse and L. Motz (eds.). *The World of the Atoms*. Vol. 1. Basic Books, New York, 1966.

8. E. A. Davis and I. J. Falconer. *J. J. Thomson and the Discovery of the Electron*. Taylor & Francis, Boca Raton, FL, 1997.

9. J. H. Freeman. Canal rays to ion implantation: 1886–1986. *Radiat. Eff.*, 100, 161–248, 1986.

10. G. E. Smith. J. J. Thomson and the electron: 1897–1899: An introduction. *Chem. Educ.*, 2(6), 1–42, 1997.

11. J. J. Thomson. On rays of positive electricity. *Philos. Mag.*, 13(77), 561–575, 1907.

12. J. J. Thomson. Ray of positive electricity. *Proc. R. Soc.*, LXXXIX, 1913.

13. G. Gamow. *The Great Physicists from Galileo to Einstein*. Dover Publications, New York, 1988. This edition is a republication of *Biography of Physics*, originally published by Harper & Brothers, New York, 1961.

14. F. Dahl. *Flash of the Cathode Rays: A History of J. J. Thomson's Electron*. Institute of Physics Publishing, Bristol, UK, 1997.

15. G. Squires. Francis Aston and the mass spectrograph. *J. Chem. Soc. Dalton Trans.*, 3893–3899, 1998.

16. J. B. Fair. A historical view of the role of ion-implantation defects in PN junction formation for devices. *Mater. Res. Soc. Symp.*, 610, B4.1.1–B.4.1.12, 2000.
17. L. Wegmann. Historical perspective and future trends for ion implantation systems. *Nucl. Instrum. Methods*, 189, 1–6, 1981.
18. P. R. Hanley. Physical limitations of ion implantation equipment. In *Proceedings of the 4th International Conference on Ion Implantation Technology, Ion Implantation: Equipment and Technique*, eds. H. Ryssel and H. Glawishnig, 2–24. Springer-Verlag, Berlin, 1983.
19. C. M. McKenna. A personal historical perspective of ion implantation equipment for semiconductor applications. In *IIT School Book (2000)*, ed. J. Ziegler, 1–45. Ion Implantation Technology Co., Chester, MD, 2000.
20. J. F. Ziegler. The history of integrated circuits and ion implantation. In *IIT School Book (2010)*, ed. J. Ziegler, 1-1–1-20. Ion Implantation Technology Co., Chester, MD, 2011.
21. L. Wegmann. *The Historical Development of Ion Implantation*, ed. J. F. Ziegler, 3–49. Academic Press, Boston, 1984.
22. C. B. Yarling. History of industrial and commercial ion implantation 1906–1978. *J. Vac. Soc. Technol. A*, 18(4), 1746–1750, 2000.
23. P. H. Rose. A history of commercial implantation. *Nucl. Instrum. Methods Phys. Res. B*, 6, 1–8, 1985.
24. P. H. Rose and G. Ryding. Concepts and designs of ion implantation equipment for semiconductor processing. *Rev. Sci. Instrum.*, 77, 111101-1–111101-12, 2006.
25. J. R. Conrad and K. Sridharan (eds.). *Papers from the First International Workshop on Plasma-Based Ion Implantation*. Published for the American Vacuum Society by the American Institute of Physics, New York, 1994.
26. S. B. Felch, M. I. Current, and N. W. Cheung. Plasma immersion ion implantation (PIII). In *IIT School Book (2010)*, ed. J. Ziegler, 4-1–4-28. Ion Implantation Technology Co., Chester, MD, 2011.
27. A. Kirkpatrick. Present status of commercial GCIB equipment and requirements for increased cluster beam current. In *Extended Abstracts of the Workshop on Cluster Ion Beam Process Technology*, October 12–13, 2000, pp. 17–18.
28. J. J. Sun, K. Shimazawa, N. Kasahara, K. Sato, T. Kagami, S. Saruki, S. Araki, and M. Matsuzaki. Magnetic tunnel junctions on magnetic shield smoothed by gas cluster ion beam. *J. Appl. Phys.*, 89, 6653–6655, 2001.

29. D. B. Fenner, J. Hautala, L. P. Allen, T. G. Tetreault, A. Al-Jibouri, J. I. Budnick, and K. S. Jones. Surface processing with gas-cluster ions to improve giant magnetoresistance films. *J. Vac. Sci. Technol. A*, 19(4), 1207–1212, 2001.
30. C. Eggs, E. Schmidhammer, and A. Schaufele. Yield enhancement for BAW production using local corrective etching. In *Proceedings of the Workshop on Cluster Ion Beam Process Technology*, November 2006, pp. 46–51.
31. J. Hautala and J. Borland. Infusion processing solutions for USJ and localized strained-Si using gas cluster ion beams. In *12th IEEE International Conference on Advanced Thermal Processing of Semiconductors—RTP2004*, 2004, pp. 37–45.
32. R. MacCrimmon, J. Hautala, M. Gwinn, and S. Sherman. Gas cluster ion beam infusion processing of semiconductors. *Nucl. Instrum. Methods Phys. Res. B*, 242, 427–430, 2006.
33. I. Yamada, J. Matsuo, and N. Toyoda. Summary of industry-academia collaboration projects on cluster ion beam process technology. In *16th International Conference on Ion Implantation Technology IIT 2008*, eds. E. G. Seebauer, S. B. Felch, A. Jain, and Y. Kondratenko, 415–418. AIP Conference Proceedings 1066. American Institute of Physics, New York, 2008.
34. J. Matsuo. Study of surface reaction dynamics by highly activated beams (in Japanese). PhD thesis, Kyoto University, April 1999.
35. N. Toyoda. Nano-processing with gas cluster ion beams, PhD thesis, Kyoto University, February 1999.
36. T. Aoki. Molecular dynamics simulation of cluster ion impact on solid surface. PhD thesis, Kyoto University, January 2000.
37. T. Seki. Nanoscale observation and analysis of damage formation and annealing processes in ion beam interactions with surface. PhD thesis, Kyoto University, January 2000.

2

History and Milestones of Cluster Beam Development

Chapter 1 briefly introduced cluster ion beam technology and equipment. Chapter 2 discusses the history of cluster beam development, the equipment used, some simulation results, and the evolution of cluster beam sources.

2.1 Cluster Beam Discovery and Early Fundamental Progress

The history of cluster beam formation dates back to the end of the nineteenth century. The historically important achievement of a cluster beam was associated with formation of supersonic gas flow through a de Laval nozzle, which had been invented and patented in 1888 by Swedish inventor Gustaf de Laval [1]. The de Laval nozzle has a converging–diverging structure that remains in broad use today. De Laval's nozzle was initially employed in the steam turbine, where it could increase the steam jet to supersonic speed. The revolution that de Laval introduced was that his steam turbine employed the kinetic energy of steam, rather than its pressure [2]. Figure 2.1 shows a schematic of de Laval's steam turbine. Figure 2.2 shows a sketch of the nozzle as shown in de Laval's

Figure 2.1 Schematic of de Laval's steam turbine, which used converging–diverging nozzles to accelerate the flow to supersonic speeds.

patent document. Since its beginning, the de Laval nozzle has been widely used in various technological applications, such as rocket engines and gas turbines, and it has become adopted for many material science purposes.

In the 1930s, John Yellott conducted research on cluster formation in terms of supersaturation and condensation of gas and vapor during supersonic expansion from a nozzle [3, 4]. Yellott made pressure measurements of the flow within a nozzle and observed that the pressure of the flow deviated from usual adiabatic expansion values. Figure 2.3 shows a sketch of one of his de Laval nozzles and shows static pressures within the nozzle. By measuring the static pressures, which were found to increase abruptly within the nozzle, Yellott was able to describe the occurrence of supersaturation and condensation in the flow. Around the same time, supersaturation phenomena in steam turbines and high-speed wind tunnels and in the formation of aerosols and fogs were being studied by many investigators [5, 6]. The investigations mentioned above are of interest because they represent the beginnings of research concerning generation of gas clusters.

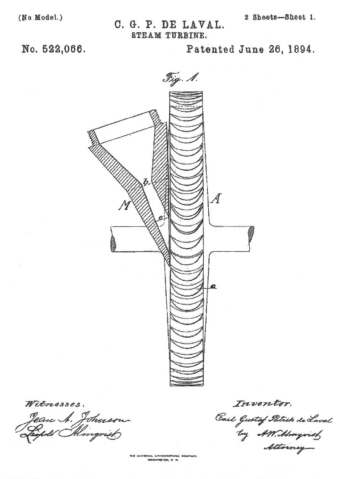

(No Model.)

C. G. P. DE LAVAL.
STEAM TURBINE.

2 Sheets—Sheet 1.

No. 522,066.

Patented June 26, 1894.

Figure 2.2 Sketch of the nozzle as shown in de Laval's patent document.

In 1951, advances in supersonic nozzle beams were made by Arthur Kantrowitz and Jerry Grey, who theoretically investigated the distributions of molecular velocities from conventional effusive sources and from nozzle sources [7]. They found that the gas velocity from the nozzle source was better monochromatized, and that the average velocity was much higher than that from the effusive source. Following up on the theoretical study and using ammonia gas, George Kistiakowsky and William Slichter provided experimental evidence of these unique characteristics [8]. Their experimental results showed that the intensity of a molecular beam of ammonia gas increased abruptly at source

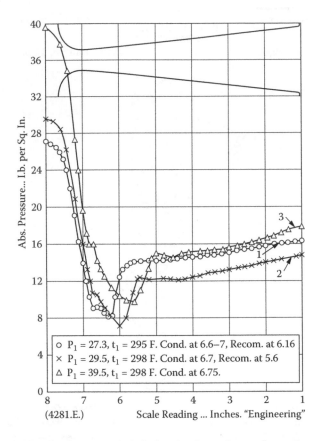

Figure 2.3 Nozzle characteristic behavior in early condensation study.

pressures beyond some critical point, as shown in Figure 2.4a. The maximum intensity observed exceeded by more than a factor of 20 that from the effusive source.

Subsequently, E. W. Becker and his coworkers made similar observations with other gases, including hydrogen, as shown in Figure 2.4b, and they explained this characteristic behavior to be associated with the onset of condensation of atoms or molecules of the beams, that is, with cluster generation [9]. Since the gas was pure, they concluded that the condensation took place spontaneously as a result of homogeneous nuclei formation.

Related investigations by Becker and coworkers were done to further demonstrate formation of clusters by velocity measurements [10]. Figure 2.5a shows construction of their experimental equipment. Figure 2.5b shows time-of-flight (TOF)

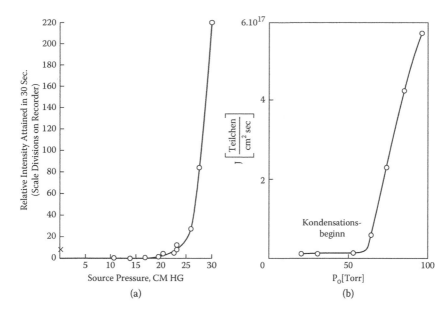

Figure 2.4 (a) Ammonia beam intensity as a function of nozzle source pressure. (b) Hydrogen beam intensity as a function of nozzle source pressure.

velocity distribution spectra obtained with cooled argon (upper) and hydrogen (lower). The spectra contain sharp peaks due to individual molecules and broad lower peaks corresponding to lower-velocity species representing molecular clusters of different sizes.

Other research to obtain high-intensity molecular gas beams was conducted by K. Bier and Otto Hagena [11] and Roger Campargue [12], who studied nozzles and skimmer structures and the influence of their configurations upon beam characteristics. Typical representative nozzles, including capillary, cylindrical hole, conical, converging, and converging–diverging types, were tested in their experiments, which included studies of the effects of distance between the nozzle and a skimmer. Detailed studies of cluster formation with respect to nozzle geometry, source gas pressure, and temperature were carried out by Hagena and Obert [13]. The fundamental concept of cluster formation has been described in many excellent review papers [14–18].

Equipment to produce highly intense cluster ion beams was first developed in Europe. According to a status report

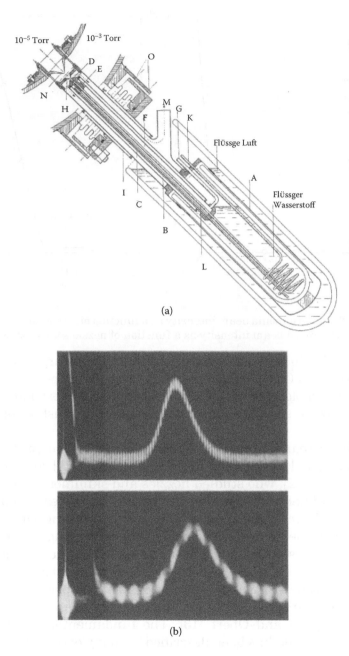

(a)

(b)

Figure 2.5 (a) Cluster beam source used by Becker and colleagues. (b) Time-of-flight spectra from hydrogen gas (top) and argon gas (bottom).

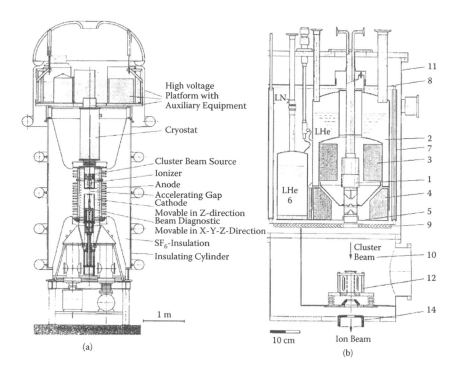

High voltage
Platform with
Auxiliary Equipment

Cryostat

Cluster Beam Source
Ionizer
Anode
Accelerating Gap
Cathode
Movable in Z-direction
Beam Diagnostic
Movable in X-Y-Z-Direction
SF$_6$-Insulation
Insulating Cylinder

1 m

(a)

LN$_2$

LHe

LHe
6

Cluster
Beam

10 cm   Ion Beam

(b)

11
8

2
7
3
1
4
5
9

10

12

14

Figure 2.6 Schematic diagram of the test assembly for (a) high-gradient acceleration of cluster ions and (b) detailed cluster beam generator.

submitted by the Institüt für Kernverfahrenstechnik (IKVT), Karlsruhe, Germany, in July 1974, it was planned that cluster ion equipment would be employed for a fusion injector. The proposal objectives called for construction of an accelerator able to deliver 10 A of hydrogen cluster ions at 1 MeV [19, 20]. A schematic diagram of the test assembly that was constructed for high-gradient acceleration of cluster ions is depicted in Figure 2.6a and b. The figure shows cross-sectional views of the vacuum chamber, which contained the cluster beam source and the ionizer. The clusters were produced by expansion of hydrogen gas at low temperature, typically 30 K, from a stagnation reservoir through a converging–diverging conical nozzle into vacuum. The beam was defined by a skimmer and a collimator. In the ionizer, shown in Figure 2.7, the beam was exposed to bombardment by transversely traveling electrons emitted from indirectly heated cathodes to an anode grid [21]. The size distribution of the hydrogen cluster ions is shown in Figure 2.8 [19]:

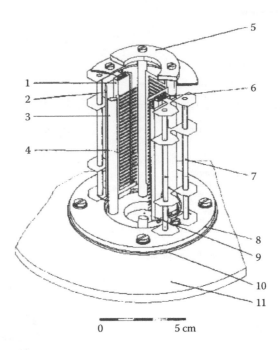

Figure 2.7 Cutaway view of an ionizer.

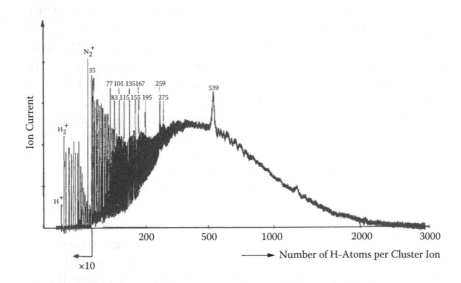

Figure 2.8 Size distribution of hydrogen cluster ion beam.

the cluster ions produced were distributed from 50 to 3000 atoms/cluster. Development of this high-energy cluster accelerator was conducted by international collaboration [22–24]. Cluster beams formed by this method would not be practical for material processing, but this original work became a base for promoting the investigation of gas cluster ions for material science surface-processing applications.

2.2 Development of Gas Cluster Beam Equipment

2.2.1 Precursor Technology

Before the development of gas cluster ion beam (GCIB) technology, researchers had attempted to identify a cluster ion technique able to take advantage of the unique characteristics that would be associated with ions comprising large numbers of atoms. During the 1970s and 1980s, a vaporized metal cluster ion beam concept known as ionized cluster beam (ICB) was extensively investigated at Kyoto University, and also elsewhere in Japan and the United States, for the purpose of forming high-quality thin films [25, 26]. The technology was originated by Toshinori Takagi of Kyoto University and was described in U.S. Patent 4152478, "Ionized-Cluster Deposited on a Substrate and Method of Depositing Ionized Cluster on a Substrate," filed October 23, 1975. Takagi was the first to recognize the considerable technical advantages that cluster ions would offer for practical processes such as film formation. The ICB concept was based on forming metal vapor clusters using Knudsen effusion cells heated to very high temperatures [27]. An ICB source shown in the original patent is illustrated in Figure 2.9. The effectiveness of cluster generation from such sources was a subject of much debate relative to the ICB concept, but excellent thin films of many materials were achieved by this technique.

During the 1980s, the exceptionally high-quality thin films produced by ICB caused the technique to be recognized as one of the pioneering approaches for ion-assisted deposition. Nevertheless, questions existed as to whether the consistently

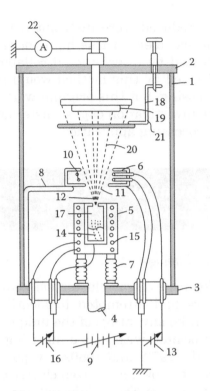

Figure 2.9 Sketch of ionized cluster ion beam apparatus shown in the first patent by Takagi.

high quality of the films deposited by ICB could actually be attributed to the role played by cluster ions or instead was due primarily to the favorable geometry of the ICB source for growth assistance by monomer ions that were also present. To address these questions and conclusively determine whether cluster ions from typical thin-film materials such as Al and Au, not from such high-vapor-pressure materials as Pb, were present in sufficient numbers in the vapor streams produced by ICB sources to allow them to be credited as a dominant factor in the film growth, a collaborative project was conducted by I. Yamada of Kyoto University, W. L. Brown of Bell Laboratories, and M. Sosnowski of New Jersey Institute of Technology. The equipment shown in Figure 2.10, consisting of an ICB source within a large ultra-high-vacuum chamber, was constructed in Japan, and in January 1989 was transferred to Bell Laboratories to be integrated with an excimer

Figure 2.10 ICB deposition equipment for cluster beam analysis that was transferred to Bell Labs in 1989.

laser apparatus designed for cluster ionization and TOF determinations of cluster size distributions.

Figure 2.11 shows (a) the schematic arrangement for the laser ionization time-of-flight measurements and (b) a typical Ag TOF spectrum obtained by experiment. From comprehensive investigations conducted using the equipment, it was concluded that the cluster ion intensities within the ICB metal vapor streams were too low to be significant to the film growth mechanisms. Detailed results of the study were published in 1991 [28, 29]. After the investigation, ICB research at Kyoto University was terminated.

Beginning at approximately the same time as the studies at Bell Laboratories, a number of researchers in the United States and Europe undertook examinations of the ICB technology. Their objective was to determine whether ICB sources could generate large cluster ions in sufficient numbers to have significant influence on the film characteristics that were being achieved. The examinations resulted in similar conclusions that conditions within the vapor streams emerging from

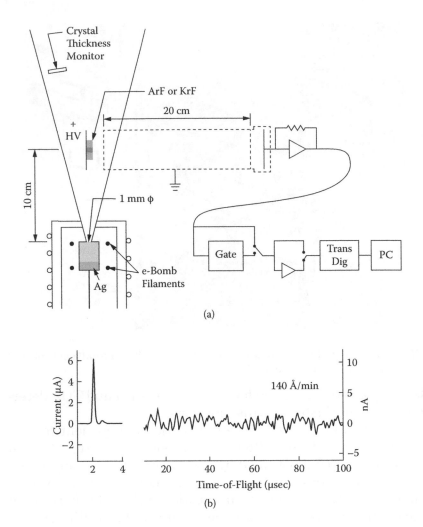

Figure 2.11 (a) Schematic of the arrangement of laser ionization time-of-flight measurements. (b) A typical TOF spectrum of vapor from a Knudsen cell containing Ag at pressure of ~10 Torr.

an ICB effusion cell source were not adequate to produce large clusters. W. Knauer [30] analyzed whether clusters were formed by homogeneous nucleation in the vapor phase during passage through the nozzle or by heterogeneous nucleation on the nozzle and crucible walls. He suggested that under conditions providing adequate supply of critical embryos within the nozzle and on the adjacent crucible wall, large clusters might form and be ejected out of the nozzle. J. G. Pruett and colleagues [31] made TOF mass spectrographs of Ge, Ag, Zn, Mg, Se, and Te vapor streams. All spectra that were obtained showed only small clusters (<10 atoms/cluster). The researchers concluded that small clusters were present in the source vapor before expansion and were ejected through the nozzle. S.-N. Mei and coworkers [32] used a retarding potential technique and found that no large Mg clusters (>100 atoms/cluster) could be detected. He concluded, and reported in 1988, that the stagnation pressure prior to the expansion was too low to allow large cluster formation. Sosnowski and coworkers [33] also came to the same conclusion based on another retarding potential method experiment. In 1990, Y. Franghiadakis and P. Tzanetakis used a TOF mass spectroscopy technique to monitor ion mass distribution from germanium [34]. Their results indicated that no clusters larger than a few atoms were observed under conditions relevant to practical applications. D. Turner and H. Shanks reported in 1991 their conclusion that an ICB source did not produce large clusters in quantities capable of affecting film growth [35]. In 1992, Hagena described his evaluation of cluster formation in terms of his Hagena parameter for free jet conditions and concluded that for typical ICB operating parameters, clusters could not form [36, 37]. J. Gspann reported in 1997 that the high densities of vapor flow required for condensation to occur in adiabatic nozzle expansions are difficult or impossible to achieve in thermally heated sources [38].

Despite the fact that the promise of large cluster ions was not accomplished with ICB, the concept attracted much attention and led to broad realization of the important advantages that cluster ions would bring to film deposition and other industrial applications if effective cluster ion sources could be produced.

2.2.2 Origination of Gas Cluster Ion Beam Technology

After ICB research at Kyoto University was discontinued in 1988, subsequent work at the Kyoto University Ion Beam Engineering Experimental Laboratory directed by Yamada became focused on cluster beam formation employing gas expansion through supersonic nozzles. The results showed that simple nozzles having converging–diverging shapes could at room temperature produce strong cluster beams from various gases. When such nozzles were incorporated into equipment with capabilities for cluster ionization and transport, the gas cluster ion beam concept emerged.

The GCIB equipment originally constructed at Kyoto University is shown in Figure 2.12a. Using this equipment,

(a)

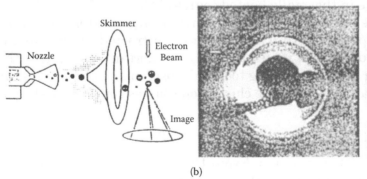

(b)

Figure 2.12 (a) Original GCIB equipment and (b) confirmation of cluster beam formation by electron diffraction pattern using an e-beam across the cluster beam. Shown are Debye–Scherrer ring patterns identifying the presence of macroparticles in the beam.

confirmation of cluster beam formation was made by observation of electron diffraction patterns that resulted from passing an e-beam through the gas stream emerging from the nozzle. These so-called Debye–Scherrer ring patterns, shown in Figure 2.12b, demonstrated the presence of macroparticles in the beam. This work then led to research and systematic development of GCIB techniques and to investigations of new ion–solid interactions produced by gas cluster ion impacts.

As work with gas cluster ions at Kyoto University progressed, several types of GCIB equipment were constructed, including, among several others, in 1992 a high-energy irradiation system able to accelerate to 200 kV [39], in 1996 an ultrahigh-vacuum (UHV) system for irradiation under 10^{-9} Torr vacuum, and in 1998 a compact GCIB system for cluster secondary ion mass spectroscopy (SIMS) and x-ray photoelectron spectroscopy (XPS) applications [40]. Figure 2.13 shows the configuration of the 200 kV GCIB equipment in which the entire gas source, nozzle, skimmer, ionizer, einzel lens and ExB mass filter, vacuum chambers, and pumps were all at 200 kV potential with the accelerated cluster ion beam delivered to targets at ground potential. Beam currents to targets were only on the order of a few picoamperes, but many important studies were conducted with the apparatus. From this equipment, it was found that beams of large gas cluster ions

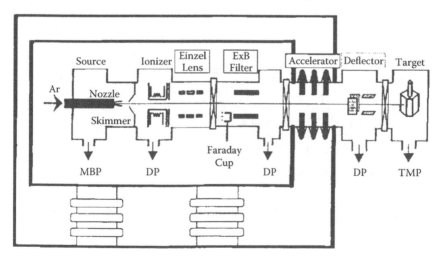

Figure 2.13 200 kV GCIB equipment.

are difficult to transport over long distances through simple drift and acceleration tubes because divergence of the beams is large compared with the situation of single-atom ion beams at the same potentials. At the time it was also not clear that the gas cluster ions would be strongly affected by collisions with residual gas atoms, which cause the ions to lose energy by fragmentation. These effects were observed and understood over time and, after a decade, have become clearly explained by experiments in recent investigations [41].

Researchers were especially interested in determining whether GCIB could be used to irradiate samples under UHV conditions. The reason was that in the formation of the gas cluster ion beam, a large amount of gas flow through the nozzle into the source chamber was necessary, which might overwhelm the operation of other vacuum chambers connected to the source chamber. In order to be certain that these concerns could be overcome, UHV-type GCIB equipment was developed [42]. Figure 2.14 shows an example of such equipment, which included an auger analysis unit, a RHEED gun for crystalline structure analysis, and a variable-temperature scanning

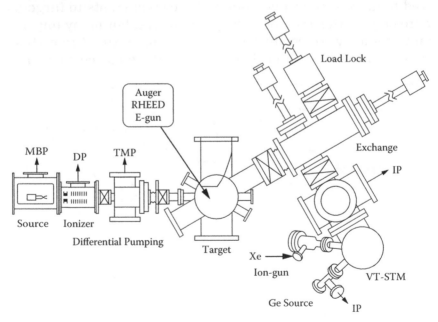

Figure 2.14 UHV-GCIB equipment.

tunneling microscope (VT-STM) for surface structure analysis. The throat diameter of the de Laval nozzle was 0.1 mm, and the skimmer hole diameter was 0.3 mm. The base pressure of the system was below 10^{-9} Torr. Under typical operation, the source chamber pressure was 10^{-3} Torr and pressure at the UHV target chamber was kept under 10^{-7} Torr by a 70 L/s turbomolecular pump. With this equipment, surface structures, defect formation, crater formation, and annealing processes were studied by using Si(111) clean surfaces.

Suitable configurations of GCIB equipment for practical industrial applications have been established, and the equipment requirements are now well known. The typical configuration of GCIB equipment is shown in Figure 2.15. A small aperture, or skimmer, transmits a primary jet core of gas clusters emerging from the expansion nozzle. Because the gas flow rate needed to produce a cluster beam is typically several hundreds of cubic centimeters per minute, a high pumping capability is needed for each vacuum chamber of a GCIB system. For fundamental research experiments, cluster beam formation can sometimes be conducted in a pulsed mode, with reduced vacuum pumping requirements, but for practical process applications, a continuous and intense cluster ion beam is required.

A typical GCIB system usually has three or four vacuum chambers: a nozzle chamber, an optional differential pumping chamber, an ionization/acceleration chamber, and a target chamber. Each chamber usually has a turbomolecular pump with a pumping speed for N_2 of 1600 to 2300 L/s. In the nozzle

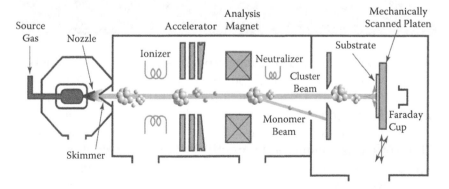

Figure 2.15 Typical configuration of GCIB equipment.

chamber, a neutral gas cluster beam is formed by passing gases at typical stagnation pressures of several atmospheres through a small expansion nozzle. Gas pressure within the nozzle chamber is typically too high to support high voltages, so the neutral cluster beam is passed through a skimmer into downstream chambers having pressures well below 1×10^{-4} Torr in order to allow ionization and acceleration. An optional differential pumping chamber can be employed to minimize background gas transport into the ionization/acceleration region where the neutral gas clusters become positively ionized by the impact of electrons accelerated from a filament. The ionized clusters are then extracted and accelerated through typical potentials of 2–30 kV by using a series of electrodes. Electrostatic lenses are utilized to focus the cluster ions, and monomer ions are filtered from the cluster ion stream by means of a strong transverse magnetic field. A neutralizer assembly injects low-energy electrons into the beam so as to minimize space charge blowup and prevent charge buildup on process targets. In the process chamber, usually the cluster ion beam is kept stationary, and material to be processed is scanned mechanically through the beam so as to obtain uniform and complete coverage. The cluster ion flux is measured by means of a Faraday cup. In discussion of GCIB equipment and processes, the acceleration voltage in kV and the cluster ion energy in keV per charge are generally used interchangeably.

2.3 Development of Gas Cluster Ion Beam Processing

In 1992, surface modification experiments using gas cluster beams were started in a collaborative study involving Yamada's group from Kyoto University, Jan Northby of the University of Rhode Island, Walter Brown of Bell Laboratories, and Marek Sosnowski of the New Jersey Institute of Technology. The work was conducted by modifying existing cluster source equipment at the University of Rhode Island. Gas cluster ion beams were created by expanding argon and helium gas mixtures through a 20-micron-diameter sonic nozzle held at low temperature (between 80K and 120K). Cluster sizes produced were in a range of 150 to 600 atoms/cluster. The results of the experiments showed that clusters of a few hundred Ar atoms having a relatively low energy of 20–30 keV created damage in a thin layer of Si while leaving little implanted Ar [43, 44]. The clusters were also used to produce surface smoothing on thin films, as had previously been observed in sputtering of Cu films by CO_2 clusters at a higher energy of 155 keV [45]. Smoothing effects on Cu, Si, and Au surfaces using low-energy accelerated Ar cluster beams (300 atoms/cluster) at 30 kV were demonstrated during these experiments. Figure 2.16 shows a scanning electron microscope (SEM) image of the surface of a Cu plate; the upper portion was irradiated by GCIB, and the lower portion was the unirradiated material. These earliest studies were the first to demonstrate that GCIB had the ability to produce unique low-energy ion–solid interactions. These interactions were subsequently found to offer many new atomic and molecular ion beam process opportunities in the areas of implantation, sputtering, and ion beam–assisted deposition.

Following the early experiments at the University of Rhode Island, extensive experimental research and molecular

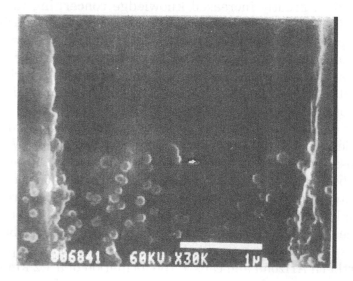

Figure 2.16 SEM picture of Cu plate surface showing smoothing produced by Ar cluster beam at 30 kV. The upper area was irradiated by GCIB, and the lower area was unprocessed.

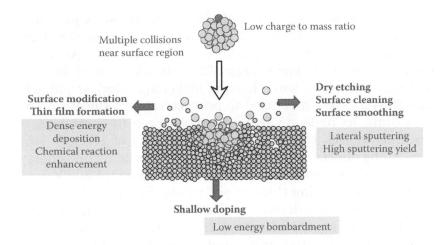

Figure 2.17 GCIB–solid surface interactions and their applications.

dynamics (MD) analytical studies of GCIB fundamental surface interactions were initiated at the Ion Beam Engineering Experimental Laboratory of Kyoto University. Investigations conducted at Kyoto University in Japan, together with Epion Corporation in the United States, for more than 10 years, resulted in greatly increased knowledge concerning interactions of energetic gas cluster ions with solid target surfaces. GCIB processing has become a powerful directional energetic chemical beam technique that offers great versatility for many surface modifications. Figure 2.17 summarizes the primary characteristics of GCIB–solid surface interactions and their areas of industrial application.

For discussion of the likely effects of cluster bombardment on solid surfaces, macroscopic analogies can be cited. When a macroscopic object impacts at high velocity upon a large rigid mass, both the surface deformation and the conversion of energy into internal degrees of freedom at the point of contact are substantial. Examples of meteor crater formation are seen on the surfaces of many planets. On the earth, approximately 50,000 years ago a metallic asteroid about 45 m in diameter impacted in northern Arizona and caused the formation of a 1.2 km wide crater. An image of this meteor crater is shown in Figure 2.18a. The shape of the crater formed by the impact is typical, with a large rim comprising ejected material [46]. Under impact

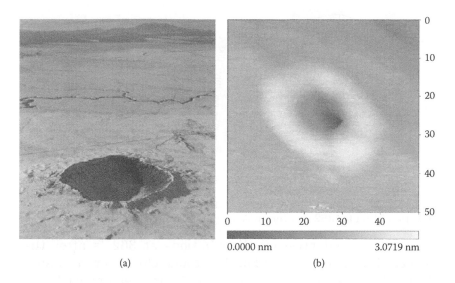

(a)   (b)

Figure 2.18 Macroscopic and microscopic views of craters: (a) meteor crater in northern Arizona (1.2 km in diameter) and (b) STM image of crater
on Au from Ar cluster ion impact (30 nm in diameter).

conditions in the hypervelocity regime (i.e., more than a few
km/s), impacted regions experience high-temperature and high-
pressure transients that produce fusion or vaporization of both
target and projectile materials and cause crater formation.

On a microscopic scale, similar craters are created on solid
surfaces as a result of impacts of high-energy particles or
heavy ions [47]. As an example, Figure 2.18b shows a scanning tunneling microscope (STM) image of a microcrater on
a Au surface due to impact of an Ar cluster ion consisting of
3000 atoms and having a total energy 150 keV. The crater
shown on the Au surface is approximately 30 nm in diameter [48], which is 4×10^{10} times smaller in diameter than the
meteorite crater shown in Figure 2.18a.

Estimations made by MD simulations indicate that interactions during energetic cluster ion impact induce transient
temperatures of tens of thousands of degrees and transient
pressures of tens of gigapascals within the impact zone of a
target surface. Equivalent phenomena are not produced by
monomer ion impacts, which involve binary collisions and do
not introduce similarly high-energy densities into the impact
volume. As examples, Figures 2.19a and b show calculated

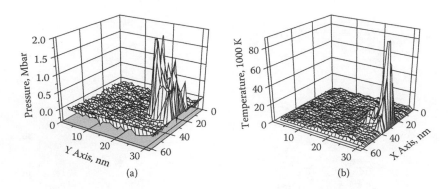

Figure 2.19 (a) Local target pressure and (b) temperature, calculated by MD simulation for 362 fs after an Ar_{349} cluster impact on a Si surface.

temperature and pressure distributions at 362 fs after the impact of a 50 keV Ar_{349} (349 Ar atoms) cluster on a Si substrate [49]. The local pressure after the cluster impact has a roughly cylindrical shape with a very sharp steep front. Such a shock wave can lead to new physical phenomena that do not occur when a single-atom ion bombards a surface. Shock waves that occur during macroscopic body impacts on planets, during explosions, during laser ablation of solid surfaces, during high-energy pulsed ion or laser beam impact on an inertial confinement fusion target, and during large organic molecule impacts are of considerable scientific and technological interest.

One of the advantages associated with the cluster ion is the effect of a very low charge-to-mass ratio. Cluster ions containing up to several thousands of atoms per charge at any given current density can transport up to thousands of times more atoms than a monomer ion beam at the same current density. For example, a 1 µA beam of cluster ions with average size of 1000 atoms/cluster can transport the same number of atoms as a 1 mA monomer ion beam.

Another advantage of GCIB for material processing is that it involves essentially low-energy individual atomic interactions even when the total energy of the clusters is high. Such interactions can be useful for producing high-rate/low-damage sputtering, shallow implantation, and other nanoscale surface modifications. From the beginning of the investigations, it was expected that clusters should produce low-energy bombardment effects. These effects were predicted by MD simulations

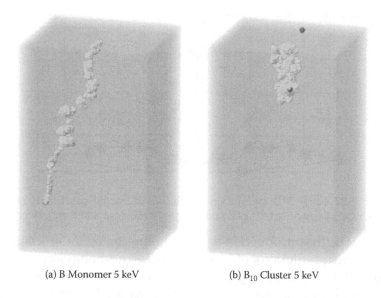

(a) B Monomer 5 keV   (b) B_{10} Cluster 5 keV

Figure 2.20 MD simulations of (a) B monomer and (b) B_{10} cluster ion implantations into Si at 5 keV.

and confirmed by experiments, for example, by comparing B monomer ion and B cluster ion implantation [50]. Figure 2.20 shows MD simulations of B monomer and cluster ion impacts into Si, illustrating the low-energy effect of cluster ion bombardment relative to monomer ion bombardment. Important differences in the range and density of the displacements produced are apparent; most notably, in the cluster ion case, the penetration range is extremely shallow and the displacements that are produced remain tightly concentrated within the impact region at the target surface. The results also show low-energy individual atomic interactions even when the total energy of the clusters was high.

Sputtering effects produced by cluster ion impacts are unique. Cluster ion bombardment produces very high sputtering yields relative to those associated with monomer ions at a similar energy. Angular distributions of ejected atoms are also considerably different. Because of the unusual directionality of the sputtered material with GCIB, the process is referred to as lateral sputtering. Lateral sputtering produces surface smoothing behavior, which normally does not occur with monomer ions. Smoothing of surfaces to atomic levels

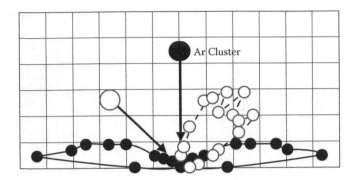

Figure 2.21 Early MD simulation prediction of lateral sputtering of gold due to normally incident 20 keV and obliquely incident 8 keV Ar_{201} impacts.

became the first production use for cluster ion beam processing. Analytical predictions of lateral sputtering behaviors associated with impacts of large energetic gas clusters emerged from MD simulation studies conducted at Kyoto University by Z. Insepov [51, 52]. Figure 2.21 shows an example of predicted polar angular distributions of gold surface atoms ejected by 20 keV Ar_{201} clusters at normal incidence and by 8 keV Ar_{201} clusters at an oblique incidence of 45°.

In order to demonstrate surface smoothing for industrial applications, detailed studies were done by employing semiconductor, insulator, and metal substrates used for device fabrication. As one example, diamond thin films deposited on Si substrates by chemical vapor deposition (CVD) were subjected to irradiation using Ar or Ar/O_2 mixed cluster ion beams. Figure 2.22 shows SEM images of CVD diamond films before and after 20 keV Ar cluster ion irradiation. The average cluster size and ion dose were 2000 atoms/cluster and 1×10^{17} ions/cm², respectively. Before irradiation, the diamond surface comprised large numbers of faceted pyramidal grains typical of polycrystalline diamond film structures. The average roughness of the initial diamond surface was 40 nm, as measured by an atomic force microscope (AFM). After irradiation, the pyramidal grain structures were no longer present, and a flat surface with an average roughness of 8 nm was observed [53].

GCIB-assisted thin-film deposition has been investigated and found to be well suited to solve problems that are often associated with other deposition techniques. Direct ion beam

(a) Before irradiation
of CVD diamond film

(b) Ar$_{2000}$ cluster ion irradiation
20 keV, 1 × 10^{17} ions/cm^2

Figure 2.22 SEM images of a CVD diamond surface (a) before and (b) after Ar cluster ion irradiation. The acceleration energy and the ion dose were 20 keV and 1 × 10^{17} ions/cm^2, respectively.

deposition and ion-assisted deposition are useful processes for depositing high-quality thin films. In some delicate materials, however, they are often subject to damage induced by their energetic ions. Very low-energy ion beams are highly desirable, but high currents of low-energy beams are difficult to produce and transport.

In GCIB-assisted deposition, vapor of some material is deposited onto a substrate while simultaneous bombardment of the growth surface by gas cluster ions is occurring. (Details are discussed in Chapter 8.) Figure 2.23 shows a schematic

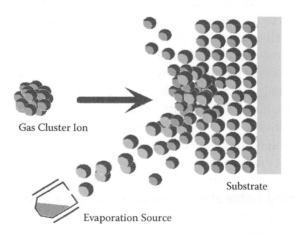

Gas Cluster Ion

Substrate

Evaporation Source

Figure 2.23 Schematic diagram of the GCIB-assisted deposition process.

diagram of the GCIB-assisted deposition process. If the cluster ions are of a nonreactive gas such as argon, this method can be used to form films having very high packing densities, smooth surfaces, improved adhesion, and so forth. If the cluster ions are of a reactive gas such as oxygen, the cluster gas atoms will react with the depositing vapor atoms. These methods have been used in the deposition of various oxide, nitride, and carbide films.

2.4 Development of Polyatomic Ion Beam Processing

In around 1991, the cluster ion beam development efforts expanded into two different, but closely related, major categories: GCIB processing and polyatomic ion beam processing. Polyatomic ion beam investigations originated during the early GCIB research in order to experimentally demonstrate the low-energy interaction effects that are associated with bombardment by multiple atom particles. Because a GCIB beam contains a wide range of cluster sizes, typically from a few hundred atoms to many thousands of atoms, the Kyoto University researchers initially had difficulty quantitatively describing the dependence of low-energy interaction effects on cluster size. In order to obtain clear experimental evidence concerning the effects of cluster size, it was desirable to use cluster ions of a single specific number of atoms per cluster, but a GCIB system equipped with a cluster size selection capability had not yet been developed. The idea then emerged to use a molecular ion consisting of a relatively large number of primary atoms of one single element. The polyatomic material candidate selected was decaborane ($B_{10}H_{14}$).

The first polyatomic ion implantation experiment was done by using equipment at the Ion Beam Engineering Experimental Laboratory of Kyoto University. Since the vapor pressure of decaborane (melting point, 99.7°C; boiling point, 213°C at standard temperature and pressure (STP) conditions) is relatively high at room temperature (4.9×10^{-2} Torr at

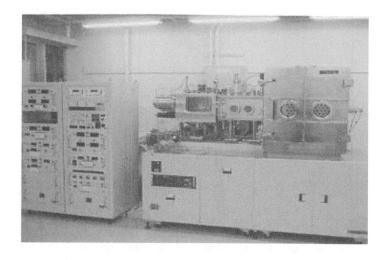

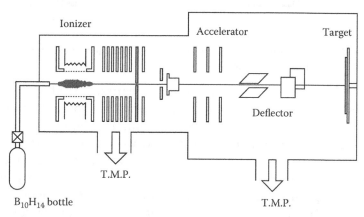

Figure 2.24 First experimental equipment used for decaborane ion implantation in 1996.

25°C), the vapor from a solid $B_{10}H_{14}$ source could be introduced directly into vacuum by using the configuration indicated in Figure 2.24.

Figure 2.25 shows mass analysis spectra of ions produced from decaborane vapor using bombardment by electrons at 20 and 40 eV. For 20 eV ionization, almost all the ionization products from the decaborane fell within a window of mass numbers from 107 to 124, indicating the presence of only $B_{10}H_x$ species. For 40 eV ionization, some fragmentation of the decaborane molecules took place.

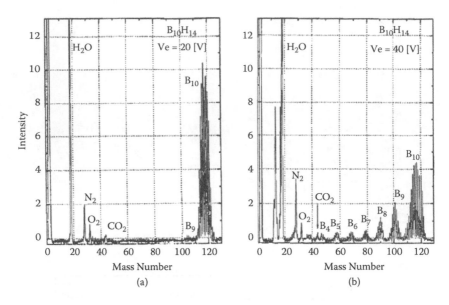

Figure 2.25 Mass analysis of decaborane ions after ionization by electrons with energies of (a) 20 eV and (b) 40 eV.

At that time the Kyoto group recognized that a decaborane source could offer important practical possibilities for low-energy-production ion implantation. The potential advantages of decaborane ion implantation were easily understood, because low-energy and high-current beam extraction were restricted in conventional ion implantation machines by basic Child–Langmuir ion beam extraction limits. Moreover, effective beam transport at very low energies was extremely difficult in implanter design because it required sophisticated beam lines involving complex electric and magnetic fields.

Implantation experiments using $B_{10}H_{14}$ were conducted. For ion implantation performed with decaborane, electron ionization potential was kept at 20 eV. The ionized $B_{10}H_{14}$ clusters were extracted and accelerated through potentials of 2 to 20 keV before being electrostatically scanned over target surfaces. It was experimentally demonstrated that the range distribution of B in Si implanted by using $B_{10}H_{14}$ at 5 keV was almost identical to that by standard B implantation at 500 eV energy. These results, which are discussed further in Section 5.2, for the first time confirmed experimentally that the effective bombarding energy of each B atom in a $B_{10}H_x$

molecular ion is roughly equal to the energy of the ion divided by the number of B atoms it contains. In the case of a GCIB cluster comprised of an elemental gas, the energy of each constituent atom is equal to the total energy of the ion divided by the number of atoms within the cluster, and consquently cluster ion beams inherently produce low-energy irradiation effects. As an example, within a 20 keV cluster ion consisting of 2000 atoms, each of the individual atoms has an energy of only 10 eV. While space charge effects make it exceptionally difficult to transport monomer ion beams at energies as low as 10 eV, equivalently low-energy ion beams can be realized by using cluster ion beams at relatively high-acceleration voltages.

Similar to what has happened throughout the history of all ion beam technology development, major progress in cluster ion beams often resulted from making the effort to attempt something radically different when it would be easy to conclude in advance that such effort would be too difficult or likely to fail. A good example was an experiment in 1996, carried out by a collaboration between Kyoto University and Kenichi Goto of Fujitsu Ltd., to fabricate a p-channel metal-oxide-semiconductor field-effect transistor (p-MOSFET) with a 40 nm gate using a $B_{10}H_{14}$ implantation for ultrashallow junction and source–drain formation [54, 55]. Figure 2.26 shows the first p-MOSFET fabricated with such an implantation. The available equipment at the Ion Beam Engineering Experimental Laboratory of Kyoto

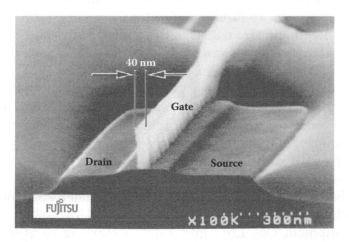

Figure 2.26 First p-MOSFET by $B_{10}H_{14}$ implantation.

University used for the experimental decaborane implantation was not qualified for semiconductor-grade clean processing. Consequently, there was a potential risk of contaminating the device fabrication equipment at Fujitsu, which would then require considerable effort and expense for cleanup. Because of the importance of the idea, however, the experiment was conducted. The successful results became the basis for a significant innovation in very low-energy ion implantation for ultrashallow junction formation and for further technological advancement. Indeed, it has evolved into an established production method for forming shallow junctions in p-MOSFETs [56]. The decaborane technology was patented in Japanese Patent H10-163123, "Ion Implantation Method and Method for Manufacturing Semiconductor Device," filed December 3, 1996, by Japan Science and Technology Agency (JST), and filed in the United States on December 6, 1996, as "Boron Doping by Decaborane."

In order to develop commercial polyatomic cluster ion implantation equipment, a JST-sponsored "risk-taking technological development project" (in Japanese, *Itaku-Kaihatsu*) was arranged. The program was selected from among candidates involving JST-held patents that originated from university research projects and where the patented technology was considered to have a high probability of being successfully developed for practical use and for promoting further research [57]. The risk-taking contract project started in 1998 at Sumitomo Eaton Nova Corporation, a subsidiary of Eaton Corporation U.S. at that time, under funding from JST. In the United States and Europe, additional experiments were conducted and were reported at the 1998 Ion Implantation Technology conference [58, 59].

The JST risk-taking project ended during fiscal year 2000, and at that time the technology was not yet fully developed. Serious concerns still existed regarding whether the aggregate atoms from molecular clusters could be adequately annealed into substitutional sites and whether end-of-range (EOR) effects and transient enhanced diffusion (TED) issues could be overcome. Further collaboration under JST support was subsequently conducted between Kyoto University, Nissin Ion Equipment Co., Ltd., and Fujitsu starting in 2003 and ending successfully in 2005.

In 2000, ion implantation equipment companies in the United States accepted the concept, and intensive development was started [60–62]. In 2001, a U.S. ion implanter manufacturing company, Axcelis Technologies (formerly named Eaton Semiconductor Equipment), announced the success of its decaborane ion implanter development program. The implanter integrated the decaborane ion source into a conventional ion implantation system. By using decaborane, a 10-fold increase in beam current was provided at the equivalent of 1/10 the ion energy [63]. A new company in the United States, SemEquip, Inc., was founded to continue the cluster ion source development for low-energy implantation that had been started at Eaton Corporation [64]. SemEquip successfully developed a cluster ion source that employed $B_{18}H_{22}$ as an alternate polyatomic source material [65].

Many other boron hydride cluster ions, such as $B_{18}H_x$ (octadecaborane) [66], $C_2B_{10}H_{12}$ (carborane) [67, 68], and $B_{36}H_{44}$ [69], were also investigated after the year 2000. In U.S. Patent 7491953 B2, "Ion Implantation Device and a Method of Semiconductor Manufacturing by the Implantation of Boron Hydride Cluster Ions," filed on December 29, 2006, Thomas Horsky and Dale Jacobson of SemEquip, Inc. discussed a broad range of potential polyatomic species, including B_nH_x, $10 < n < 100, 0 \leq x \leq n + 4$. The advantages of using such polyatomic ions for low-energy ion implantation applications were discussed at the 2010 International Workshop on Junction Technology [70], and the improvement in the productivity of cluster ion implantation was also reported [71].

References

1. C. G. P. de Laval. Steam turbine. U.S. Patent 522066, June 26, 1894, filed May 1, 1889, serial no. 309209. Patented in Belgium, September 29, 1888, no. 83196. Patented in England, April 29, 1889, no. 7143.
2. W. B. Browere Jr. *A Primer in Fluid Mechanics*. CRC Press, Boca Raton, FL, 1999. C. A. Parsons. *The Steam Turbine: The REDE Lecture*. Printed by John Clay at Cambridge Press, Cambridge, 1911.

3. J. I. Yellott. Superaturated steam. *Engineering*, March 9, 1934, pp. 303–305.

4. J. I. Yellott and C. K. Holland. The condensation of flowing steam: Condensation in diverging nozzles. *Engineering*, June 4, 1937, pp. 703–705.

5. J. V. Charyk. Condensation phenomena in supersonic flows. PhD thesis, California Institute of Technology, August 1946.

6. R. M. Head. Investigations of spontaneous condensation phenomena. PhD thesis, California Institute of Technology, August 1949.

7. A. Kantrowitz and J. Grey. A high intensity source for the molecular beam: Part I: Theoretical. *Rev. Sci. Instrum.*, 22(5), 328–332, 1951.

8. G. B. Kistiakowsky and W. P. Slichter. A high intensity source for the molecular beam: Part II: Experimental. *Rev. Sci. Instrum.*, 22(5), 333–337, 1951.

9. E. W. Becker, K. Bier, and W. Henkes. Strahlen aus Kondensierten Atomen und Molekeln im Hochvakuum. *Z. Phys.*, 146, 333–338, 1956.

10. E. W. Becker und W. Henkes. Geschwindigkeitsanalyse von Laval-Strahlen. *Z. Phys.*, 146, 320–332, 1956.

11. K. Bier and O. F. Hagena. Molecular beams. In *Rarefied Gas Dynamics*, ed. J. H. Leeuw, 260–278. Vol. II. Academic Press, Boston, 1966.

12. R. Campargue. High intensity supersonic molecular beam apparatus. In *Rarefied Gas Dynamics*, ed. J. H. Leeuw, 279–298. Vol. II. Academic Press, Boston, 1966.

13. O. F. Hagena and W. Obert. Cluster formation in expanding supersonic jets: Effect of pressure, temperature, nozzle size, and test gas. *J. Chem. Phys.*, 56(5), 1793–1802, 1972.

14. P. P. Wegener and A. A. Pouring. Experiments on condensation of water vapor by homogeneous nucleation in nozzles. *Phys. Fluids*, 7(3), 352–361, 1964.

15. P. P. Wegener. Gas dynamics of expansion flows with condensation, and homogeneous nucleation of water vapor. In *Gas Dynamics*, ed. P. P. Wegener, 163–243. Vol. 1, *Nonequilibrium Flows*. Marcel Dekker, New York, 1969.

16. P. G. Hill. Condensation of water vapour during supersonic expansion in nozzles. *J. Fluid Mech.*, 25(Part 3), 593–620, 1966.

17. G. S. Springer. Homogeneous nucleation. In *Advances in Heat Transfer*, eds. T. F. Irrine Jr. and J. P. Hartnett, 281–346. Vol. 14. Academic Press, New York, 1978.

18. P. P. Wegener and J.-Y. Parlange. Condensation by homogeneous nucleation in the vapor phase. *Naturwissenshaften*, 57(11), 525–533, 1970.

19. E. W. Becker, H. Falter, O. F. Haggena, W. Henkes, R. Klingelhofer, K. Korting, F. Mikosch, H. Moser, W. Obert, and J. Wust. Development and construction of an injector using hydrogen cluster ions for nuclear fusion devices status report as of December 1973. KFK 2016. Gesellshaft fur Kernforschung M.B.H Karlsruhe, July 1974.

20. O. F. Hagena, W. Henkes, and U. Pfeiffer. Formation and detection of high-energy cluster beams. Presented at 10th Rarefied Gas Dynamics Conference Snowmass, Aspen, CO, July 1976.

21. W. Henkes, V. Hoffman, and F. Mikosch. Ionizser for cluster ion accelerators. *Rev. Sci. Instrum.*, 48(6), 675–661, 1977.

22. H. O. Moser, J. Martin, and R. Salin. An electrostatic high-gradient accelerator for hydrogen cluster ions. *J. Phys.*, 38(7), C2-215–C2-218, 1977.

23. M. J. Gaillard, A. Schempp, H. O. Moser, H. Deitinghoff, R. Genre, G. Hadinger, A. Kipper, J. Madlung, and J. Martin. First high-energy hydrogen cluster beams: A new facility at IPN Lyon, France. Presented at the 6th International Symposium on Small Particles and Inorganic Clusters (ISSPIC 6), Chicago, September 15–22, 1992.

24. A. Schempp. A variable energy RFQ for the acceleration of heavy clusters. *Nucl. Instrum. Methods Phys. Res. B*, 88, 16–20, 1994.

25. T. Takagi. Ionized cluster beam (ICB) deposition and processes. *Pure Appl. Chem.*, 60(5), 781–794, 1988.

26. T. Takagi. *Ionized cluster beam deposition and epitaxy*. Noyes, Park Ridge, NJ, 1988.

27. T. Takagi, I. Yamada, M. Kunori, and S. Kobiyama. Vapourized metal cluster ion source for ion plating. In *Proceedings of the Second International Conference on Ion Sources*, Vienna, Austria, September 11–15, 1972, pp. 790–795.

28. W. L. Brown, M. F. Jarrold, R. L. McEachern, M. Sosnowski, G. Takaoka, H. Usui, and I. Yamada. Ion cluster beam deposition of thin films. *Nucl. Instrum. Methods Phys. Res. B*, 59/60, 182–189, 1991.

29. R. L. McEachern, W. L. Brown, M. F. Jarrold, M. Sosnowski, G. Takaoka, H. Usui, and I. Yamada. An investigation of cluster formation in an ionized cluster beam deposition source. *J. Vac. Sci. Technol. A*, 9(6), 3105–3112, 1991.

30. W. Knauer. Formation of large metal clusters by surface nucleation. *J. Appl. Phys.*, 62, 841–851, 1987.

31. J. G. Pruett, H. Windischmann, M. L. Nicholas, and P. S. Lampard. Cluster size and temperature measurement in a pure vapor source expansion. *J. Appl. Phys.*, 64(5), 2271–2278, 1988.

32. S.-N. Mei, S.-N. Yang, J. Wong, C.-H. Choi, and Y.-M. Lu. On the metal cluster formation in ionized cluster beam deposition. *J. Crystal Growth*, 87, 375–364, 1988.

33. M. Sosnowski, S. Krommenhoek, J. Sheen, and R. H. Cornely. Study of the properties of Ga beam from the nozzle source. *J. Vac. Sci. Technol. A*, 8(3), 1458–1464, 1990.

34. Y. Franghiadakis and P. Tzanetakis. Time of flight mass spectroscopy of ionized cluster beam in film deposition conditions. *J. Appl. Phys.*, 68, 2433–2438, 1990.

35. D. Turner and H. Shanks. Experimental and computational analysis of ionized cluster beam. *J. Appl. Phys.*, 70, 5385–5400, 1991.

36. O. F. Hagena. Cluster ion sources (invited). *Rev. Sci. Instrum.*, 63(4), 2374–2379, 1992.

37. P. Gatz and O. F. Hagena. Cluster beam deposition: Optimization of the cluster beam source. *J. Vac. Sci. Technol. A*, 13(4), 2128–2132, 1995.

38. J. Gspann. Reactive accelerated cluster erosion for microstructuring. In *Similarities and Differences between Atomic Nuclei and Clusters*, ed. Y. Abe, I. Arai, K. M. Lee, and K. Yobana, 299–309. AIP Conference Proceedings 416. American Institute of Physics, New York, 1997.

39. D. Takeuchi. Non-linear processes in cluster ion implantation on solid surface. PhD thesis, Kyoto University, January 1997.

40. N. Toyoda, J. Matsuo, T. Aoki, S. Chiba, I. Yamada, D. B. Fenner, and R. Tori. Secondary ion mass spectrometry with gas cluster ion beams. *Mater. Res. Soc. Proc.*, 647, O5.1.1–O5.1.6, 2001.

41. D. R. Swenson. Measurement of averages of charge, energy and mass of large, multiply charged cluster ions colliding with atoms. *Nucl. Instrum. Methods Phys. Res. B*, 222, 61–67, 2004.

42. T. Seki. Nanoscale observation and analysis of damage formation and annealing processes in ion beam interactions with surface. PhD thesis, Kyoto University, January 2000.

43. J. A. Northby, T. Jiang, G. H. Takaoka, I. Yamada, W. L. Brown, and M. Sosnowski. A method and apparatus for surface modification by gas-cluster ion impact. *Nucl. Instrum. Methods Phys. Res. B*, 74, 336–340, 1993.

44. I. Yamada, M. Ishii, Y. Yamashita, J. A. Northby, T. Jiang, C. Kim, W. L. Brown, and M. Sosnowski. Experiment on Ar cluster beam bombardment at Rhode Island University and Bell Labs. Internal report. Ion Beam Engineering Experimental Laboratory, Kyoto University, 1992.
45. P. R. W. Henkes and R. Klingelhofer. Micromachining with cluster ions. *Vacuum*, 39(6), 541–542, 1989.
46. T. Gehrels. Collisions with comets and asteroids. *Scientific American*, March 1996, pp. 54–59.
47. K. L. Merkle and W. Jager. Direct observation of spike effects in heavy ion sputtering. *Phil. Mag. A*, 44, 741–762, 1981.
48. D. Takeuchi, K. Fukushima, J. Matsuo, and I. Yamada. Study of Ar cluster ion bombardment of a sapphire surface. *Nucl. Instrum. Methods B*, 121, 493–497, 1997.
49. Z. Insepov and I. Yamada. Molecular dynamics study of shock wave generation by cluster impact on solid targets. *Nucl. Instrum. Methods B*, 112, 16–22, 1996.
50. I. Yamada and J. Matsuo. Cluster ion beam processing. *Mater. Sci. Semicond. Process.*, 1, 27–41, 1998.
51. Z. Insepov, M. Sosnowski, and I. Yamada. Molecular-dynamics simulation of metal surface sputtering by energetic rare-gas cluster impact. In *Laser and Ion Beam Modification of Materials*, eds. I. Yamada, H. Ishiwara, E. Kamijo, C. W. Allen, and C. W. White, 111–118. Vol. 17. Elsevier Science, Amsterdam, 1994.
52. Z. Insepov and I. Yamada. Molecular-dynamics simulation of cluster ion bombardment of solid surfaces. *Nucl. Instrum. Methods B*, 99, 248–252, 1995.
53. A. Yoshida, M. Deguchi, K. Kitabatake, T. Hirao, J. Matsuo, N. Totoda, and I. Yamada. Atomic level smoothing of CVD diamond films by gas cluster ion beam etching. *Nucl. Instrum. Methods B*, 112, 248–251, 1996.
54. K. Goto, J. Matsuo, T. Sugii, H. Minakata, I. Yamada, and T. Hisatsugu. Novel shallow junction technology using decaborane ($B_{10}H_{14}$). *IEDM Tech. Dig.*, 435–438, 1996.
55. D. Takeuchi, N. Shimada, J. Matsuo, and I. Yamada. Shallow junction formation by polyatomic cluster ion implantation. In *Proceedings of the 11th International Conference on Ion Implantation Technology—IIT '96*, vol. 1, issue 1, pp. 772–775. Institute of Electrical and Electronics Engineers, Piscataway, NJ, 1996.

56. K. Goto, J. Matsuo, T. Y. Tada, T. Tanaka, Y. Momiyama, T. Sugii, and I. Yamada. A high performance 50 nm PMOSFET using decaborane ($B_{10}H_{14}$) ion implantation and 2-step activation annealing process. *IEDM Tech Dig.*, 18.4.1–18.4.4, 1997.

57. K. Goto, M. Kase, J. Matsuo, I. Yamada, D. Takeuchi, N. Yoda, and N. Shimada. Boron doping by decaborane. U.S. Patent 6013332, January 11, 2000, filed December 6, 1996.

58. A. G. Dirks, P. H. L. Bancken, and J. Politiek. Low-energy implantations of decaborene (B_{10} H_{14}) ion clusters in silicon wafers. In *1998 International Conference on Ion Implantation Technology Proceedings*, Kyoto, Japan, June 22–26, 1998, pp. 1167–1170.

59. A. Agarwal, H.-J. Gossmann, D. C. Jacobson, L. Pelaz, D. Eaglesham, M. Sosnowski, J. M. Poate, I. Yamada, J. Matsuo, and T. E. Haynes. Enhanced diffusion from decaborane molecular ion implantation. In *1998 International Conference on Ion Implantation Technology Abstracts*, Kyoto, Japan, June 22–26, 1998, pp. 1–4.

60. M. C. Vella, R. Tysinger, M. Reilly, and B. Brown. Decaborane ion source demonstration. In *2000 International Conference on Ion Implantation Technology Proceedings*, Alpbach, Austria, September 17–22, 2000, pp. 527–529.

61. A. S. Perel, W. Krull, D. Hoglund, K. Jackson, and T. Horsky. Decaborane ion implantation. In *2000 International Conference on Ion Implantation Technology Proceedings*, Alpbach, Austria, September 17–22, 2000, pp. 304–307.

62. D. C. Jacobson, K. Bourdelle, H.-J. Gossmann, M. Sosnowski, M. A. Albano, V. Babaram. J. M. Poate, A. Agarwas, A. Perel, and T. Horsky. Decaborane, and alternative approach to ultra low energy ion implantation. In *2000 International Conference on Ion Implantation Technology Proceedings*, Alpbach, Austria, September 17–22, 2000, pp. 300–303.

63. Axcelis integrates decaborane technology onto ion implantation system. *Electronic News—EDN*, March 5, 2001.

64. T. N. Horsky. Decaborane vaporizer. U.S. Patent 6107634, August 22, 2000, filed April 30, 1998.

65. T. N. Horsky and D. C. Jacobson. Ion implantation device and a method of semiconductor manufacturing by the implantation of boron hydride cluster ions. U.S. Patent 7491953B2, February 17, 2009, filed December 29, 2007.

66. M. Tanjyo, T. Nagayama, N. Hamamoto, S. Umisedo, Y. Koga, N. Maehara, T. Matsumoto, N. Nagai, F. Ootuka, A. Katakami, K. Shirai, T. Watanabe, H. Nakata, M. Ktajima, T. Aoyama,

T. Eimori, Y. Nara, Y. Ohji, K. K. Saker, W. Krull, D. Jacobson, and T. Horsky. Cluster ion implantation for beyond 45 nm node novel device A applications. In *Extended Abstracts of the 8th International Workshop on Junction Technology*, 55–57. Institute of Electrical and Electronics Engineers, Piscataway, NJ, 2008.

67. A. Renau. A better approach to molecular implantation. In *Extended Abstracts of the 7th International Workshop on Junction Technology*, 107–112. Institute of Electrical and Electronics Engineers, Piscataway, NJ, 2007.

68. A. Renau. Device performance and yield: A new focus for ion implantation. In *Extended Abstracts of the 10th International Workshop on Junction Technology*, 1–6. Institute of Electrical and Electronics Engineers, Piscataway, NJ, 2010.

69. W. Krull. Advances in molecular implant technology. In *9th International Workshop on Junction Technology*, 1–5. Institute of Electrical and Electronics Engineers, Piscataway, NJ, 2010.

70. S. Qin, Y. Jeff Hu, and A. McTeer. Advanced boron-based ultra-low energy doping techniques on ultra-shallow junction fabrications. In *10th International Workshop on Junction Technology*, 255–260. Institute of Electrical and Electronics Engineers, Piscataway, NJ, 2010.

71. M. Tanjyo, N. Hamamoto, S. Umisedo, Y. Koga, H. Une, N. Maehara, Y. Kawamura, Y. Hashino, Y. Nakashima, M. Hashimoto, T. Nagayama, H. Onoda, N. Nagai, T. N. Horsky, S. K. Hahto, and D. C. Jacobson. Improvement of productivity by cluster ion implanter: CLARIS. In *10th International Workshop on Junction Technology*, 110–113. Institute of Electrical and Electronics Engineers, Piscataway, NJ, 2010.

3

Development of Cluster Beam Sources for Solid Materials

Interest in the evolution of physical properties from atoms to bulk materials is a growing area of materials science. Because clusters of various sizes could represent transitional states from atoms or molecules to solids, they became powerful investigative tools. Experimental techniques for production of clusters comprising metallic, semiconducting, and insulating atoms in various numbers have progressed in recent decades, bringing deep understanding of the differences and similarities of materials' evolution [1], such as when does a metal become a metal, and when does a cluster mimic the structure of a bulk solid? Answers to such questions have interested even those who are not scientists. For these questions, however, there was not a single answer, because various materials exhibit different evolutions of their properties with size [2]. Advances in these areas of research contribute not only to fundamental understanding, but also to many areas of technical applications, such as in chemical engineering, optical, magnetic, and electrical device fabrications. Many technical review papers and books have been published on this subject [3–6].

In this section, historically important contributions related to the development of cluster beams from solid materials are summarized. Cluster beams of metals, semiconductors, and insulating materials are formed mostly by either gas condensation or expansion of their pure vapors. Methods of vaporization of solid materials include arc discharge, thermal evaporation, and laser evaporation. These methods are usually combined with a supersonic expansion nozzle system. Direct methods for formation of clusters by sputtering and by intense laser ablation are often used for high-temperature materials. Cluster beams of large polyatomic molecules are usually produced by simple heating in ovens at temperatures low enough to avoid molecular dissociation. Various available sources for producing solid material clusters are classified as follows:

1. Gas aggregation sources
2. Pure expansion nozzle sources
3. Arc vaporization sources
4. Laser beam irradiation sources
5. Ion and plasma sputtering sources
6. Polyatomic sources

Table 3.1 lists various cluster sources, the methods they employ, and examples of typical compatible materials.

3.1 Gas Condensation and Aggregation Sources

In a gas aggregation source, vapor evaporated from an oven is condensed into clusters by mixing with an inert carrier gas. In 1930, A. H. Pfund demonstrated that a black film, termed *bismus-brack*, was formed by evaporation of bismuth in a very low-pressure bell jar. Measurement of particle size was not done at that time because of the limit of resolution of available microscopes [7, 8]. Later, in the 1940s, fine metal particles now recognized as having been clusters were produced by Ryozi Uyeda using evaporation of metals in gas at low pressures [9, 10]. Figure 3.1 shows the apparatus. Initially, zinc was heated in a tungsten wire basket in air at reduced pressure of about 1 Torr. The evaporated vapor from the basket was condensed into particles by collisions with the air. The particles

TABLE 3.1
Various Types of Cluster Sources, Their Methods of Cluster Formation, and Representative Cluster Materials

Cluster Sources	Method	Example Cluster Materials
Gas condensation and aggregation sources	Thermal evaporation of solid material in a carrier gas with condensation by gas collisions	Sb, Bi, Pb, Ag, In
Pure expansion sources	Thermal evaporation of a high-vapor-pressure solid material through an expansion nozzle without an added carrier gas	Cs
Arc vaporization sources	Arc discharge evaporation of solid material in a carrier gas with condensation by collisions	C, W, Ag
Laser irradiation sources	Laser beam evaporation of solid material in a carrier gas with condensation by collisions	Cu, Ag, Ni, Fe, Si, Mo, W
Ion and plasma sputtering sources	Injection of solid material into a carrier gas by ion or plasma sputtering with condensation by collisions	Ag, Cu, Mo
Polyatomic sources	Direct evaporation of polyatomic materials	$B_{10}H_{14}$, $B_{18}H_{24}$, $C_2B_{10}H_{22}$, $C_{14}H_{14}$, $C_{16}H_{10}$, C_7H_7

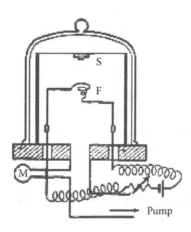

Figure 3.1 Evaporator for fine metal particle generation.

were collected on a clean brass plate. Then, by measurements using electron diffraction, the particle size and crystalline structure were determined. The average size of the clusters formed was 10–200 Å, depending on the vapor pressure and temperature of the evaporated material. Particles were subsequently also produced from other metals, including Mg, Al, Cr, Mn, Fe, Co, Ni, Cu, Ag, Cd, Sn, Au, and Pb; these were also studied by electron diffraction. The species prepared in this manner were clusters, but they were not cluster beams.

In 1978, Gilbert Stein and coworkers reported a supersonic metal cluster source combined with an evaporation oven [11]. At the exit of the nozzle an electron beam system was incorporated in order to observe electron diffraction from the emerging vapors. Figure 3.2a shows the configuration of the apparatus. The metal was vaporized from an oven in the presence of Ar, or another carrier gas, which initiated the metal nucleation and growth in the gas phase, near the evaporating surface. The evaporation oven was operated with a tungsten filament. The gas supply and the pumping speed within the annular region between the orifices could be controlled so as to vary the pressure in the oven in order to alter the jet velocity, which in turn controlled the gas mixing time. The crystalline structures and diameters of various metal clusters were investigated. Figure 3.2b shows the typical electron diffraction patterns obtained from Pb. For curves A, B, and C, Ar source gas pressures were 0.83, 0.67, and 0.55 Torr, respectively [12]. Corresponding Pb cluster diameters were found to be 82, 60, and 40 Å. It was clearly shown that the crystalline structures changed according to the operational conditions. Various other evaporated metals (Ag, In, and Bi) in various gases (He, Ar, CO_2, and SF_6) were also examined, and resulting cluster diameters were found to range from 30 to 110 Å [13].

Figure 3.3 shows experimentally obtained cluster size (D_e) dependencies on the product of carrier gas pressure (P_0) times the temperature of the evaporating metal sample (T_{0m}). An increase in average size with oven temperature or carrier gas pressure was apparent. This was explained to have resulted because increasing the carrier gas pressure enhanced the heat transfer while inhibiting the mass transport. Ag cluster sizes were different when argon and helium were used. The

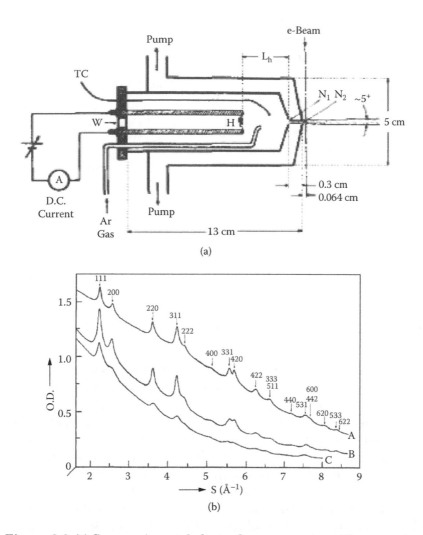

Figure 3.2 (a) Supersonic metal cluster beam generator with evaporator and (b) electron diffraction patterns of lead clusters.

differences were considered to reflect their differing atomic masses and different collisional cross sections. A larger number of small Ag clusters were produced with helium gas because of more rapid cooling of the hot metal vapor. The chemically inert molecular gas sulfur hexafluoride, SF_6, with its many internal degrees of freedom (i.e., large heat capacity per molecule), was found to be more effective in the early stages of forming large clusters.

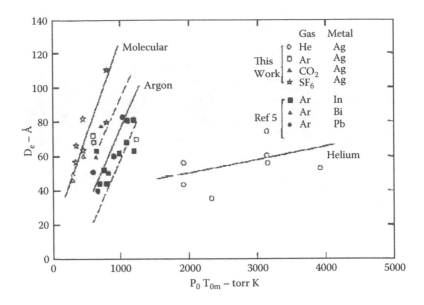

Figure 3.3 Cluster sizes determined by electron diffraction as a function of oven pressure (P_0) and temperature (T_{0m}).

Beams of metal clusters of Sb, Bi, and Pb containing from two to several hundred atoms per cluster were reported by Klaus Sattler and coworkers in 1980 [14]. Figure 3.4a shows the cluster source. Intensities of the cluster beams produced with inert gas flows at pressures up to 20 mbar for condensation were about 1000 times higher than had been achieved by an atomic beam without inert gas condensation. A time-of-flight (TOF) mass spectrum of Sb clusters produced by condensation in a He atmosphere is shown in Figure 3.4b. Sattler showed clusters of sizes of up to 100 atoms for Sb, 200 atoms for Bi, and 400 atoms for Pb created by this source. In order to produce binary metal alloy clusters, a system using two of the sources shown in Figure 3.4a was also employed [15].

Two types of metal cluster sources, a nozzle source and an evaporation source, were reported by A. E. T. Kuiper and coworkers in 1981 for thin-film deposition [16]. The first nozzle-type source consisted of an oven and a supersonic nozzle, as shown in the schematic diagram in Figure 3.5a. Clusters of Ag were produced by evaporation from a molybdenum boat in combination with adiabatic expansion of He. The chamber could

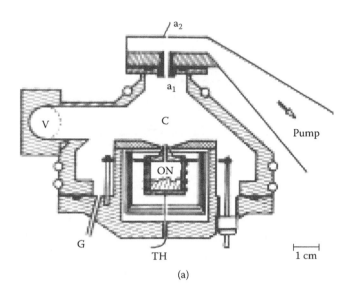

(a)

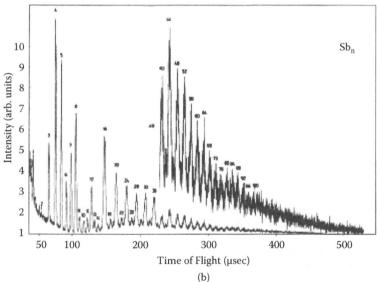

(b)

Figure 3.4 (a) Gas aggregation cluster source and (b) example TOF mass spectrum of Sb clusters.

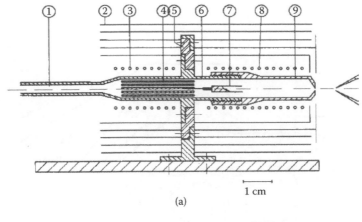

1 cm

(a)

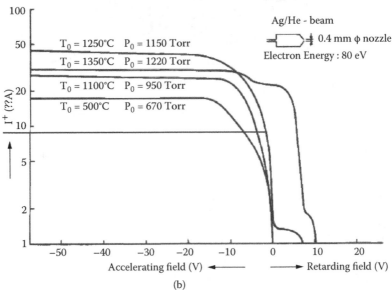

Ag/He - beam

0.4 mm ϕ nozzle

Electron Energy : 80 eV

$T_0 = 1250°C$ $P_0 = 1150$ Torr

$T_0 = 1350°C$ $P_0 = 1220$ Torr

$T_0 = 1100°C$ $P_0 = 950$ Torr

$T_0 = 500°C$ $P_0 = 670$ Torr

I^+ (??A)

Accelerating field (V) ◄———   ———► Retarding field (V)

(b)

Figure 3.5 (a) Schematic diagram of the nozzle source for metal clusters and (b) retarding field analyses of silver nozzle beams.

be heated to approximately 1500°C by an external tungsten heating coil (8). Carrier gas entered the reservoir via a heat exchanger (4) formed by six narrow channels. This section was also heated externally by a second tungsten coil (3), so that the temperature of the carrier gas could be adjusted more or less independently of the reservoir temperature. Cluster sizes were measured by means of a retarding field technique that gave results as shown in Figure 3.5b. At a source temperature of 1350°C, and with carrier gas pressures of 1150–1220 Torr, mean cluster sizes were typically about 25 atoms.

The second metal source, the evaporation source, was constructed to have almost the same structure as that of Stein [11]. With this source, large Ag and Ge metal clusters with 10^4–10^5 atoms were produced by proper choices of source pressures and orifice diameters. The cluster beams were ionized by bombardment with 300 eV electrons at emission currents of 50–100 mA and were accelerated by potentials of up to 5 kV for thin-film deposition.

A source for producing high-intensity beams of size-selected metal clusters was reported by Ian Goldby in 1996 [17]. The source, which is shown in Figure 3.6, was used to investigate thin-film deposition processes. The equipment consisted of condensation chamber (a), helium inlet tube (b), crucible (c), water cooling (d), nozzle (e), vacuum pump (f), ionizer (g), and skimmer (h). For cluster beam ionization, a hot-cathode plasma ionizer was used. Construction of the plasma ionizer is shown in Figure 3.7 [18]. The U-shaped filament was heated by an electric current and was biased at between 70 and 120 V negative with respect to the nozzle and the skimmer. Seven samarium–cobalt magnets were arranged in a circle around the nozzle in order to confine the plasma and

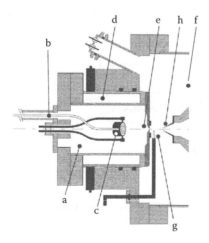

Figure 3.6 Schematic of the cluster source.

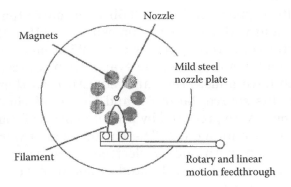

Figure 3.7 Sketch of hot-cathode plasma ionizer.

greatly increase the path length of electrons passing through the plasma. Ionization enhancement was achieved because the magnetic field forced electrons and small-mass ions to follow helical paths around the field lines, thereby greatly increasing their interaction lengths, and thus the probability of ionizing collisions with neutral clusters. The estimated ionization efficiency was nearly 20%, higher than would be expected with standard electron bombardment-type ionization.

Ag cluster mass spectra as functions of crucible temperature are shown in Figure 3.8. In this experiment, a conical nozzle

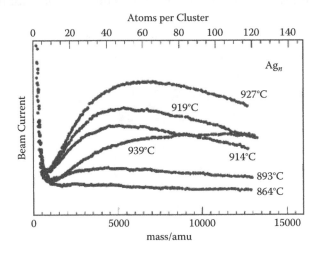

Figure 3.8 Silver cluster mass spectra as a function of crucible temperature.

with 0.8 mm throat diameter was used. Typical pressures ranged from 1 to 20 mbar. As the temperature was increased, small clusters were produced initially, and then larger clusters started to form. Cluster sizes were distributed from 10 to 120 atoms/cluster.

3.2 Pure Expansion Source

In 1991, Gspann reported the generation of large Cesium clusters with up to 2500 atoms/cluster using nozzle expansion of pure metal vapor without carrier gas [19]. Figure 3.9 shows the experimental equipment, which consisted of a cluster source with ionization and acceleration devices. Cesium vapor was expanded from either of two types of converging–diverging nozzles used for this experiment: an axisymmetric nozzle and a square cross section nozzle. Cluster beams were partially ionized by means of electron pulses, and the ionized clusters were collected at a distance of several centimeters away from the ionizer. Acceleration potential was made either negative or positive in order to detect positive and negative clusters. A surprising result described by Gspann was that even when the acceleration voltage was zero, both positive and negative ion signals were detected. Figure 3.10 shows the cluster size dependence upon ionizing electron energy.

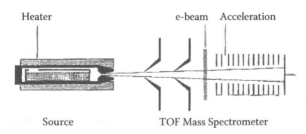

Figure 3.9 Schematic view of the pure expansion cesium cluster beam setup.

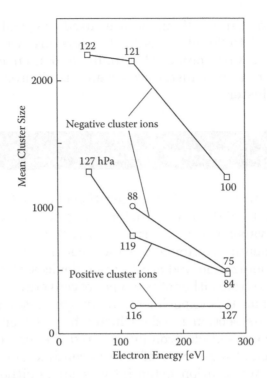

Figure 3.10 Mean cluster sizes in atoms per elementary charge versus energy of the ionizing electron for two types of nozzles: axisymmetric nozzle (circles), and square cross section nozzle (squares). Numbers in the figure show vapor pressures corresponding to the crucible temperatures.

3.3 Arc Vaporization Sources

A pulsed arc cluster ion source (PACIS) was reported by G. Ganteför and colleagues in 1990 [20]. As shown in Figure 3.11, two metal rods were mounted opposite each other in a ceramic block. A pulsed arc discharge was produced between the rods by applying high voltage from a thyristor power supply. A pulsed carrier gas valve was attached to the ceramic block, and the gas flushed the resulting plasma through a channel into a conical nozzle. Various nozzles and channel geometries were tested. In these sources, no additional ionizing agent was needed in order to charge the clusters. However, the cluster

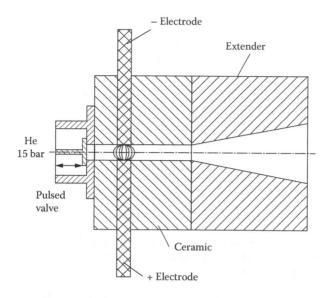

Figure 3.11 Pulsed arc cluster ion source.

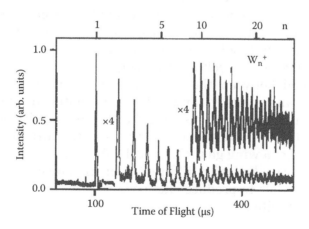

Figure 3.12 Time-of-flight mass spectrum of tungsten cluster ions.

ions produced were positively as well as negatively charged. Figure 3.12 shows the mass spectrum of positive W clusters, which were found to have sizes of up to about 25 atoms.

Paulo Milani and coworkers reported a highly intense pulsed microplasma cluster source (PMCS) in 1999 [21]. The source achieved efficient vaporization of metal or carbon source materials by means of a pulsed plasma discharge between

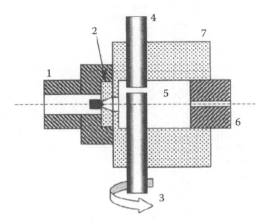

Figure 3.13 Cross section of the PMCS source: (1) pulsed valve, (2) insulating valve nozzle, (3) cathode, (4) anode, (5) thermalization cavity, (6) cylindrical nozzle, and (7) ceramic main body.

the source material electrodes located in a gas flow channel. Figure 3.13 shows the schematic of the PMCS. Under standard repetitive pulse mode operation, He gas flowed past a solenoid valve, through the discharge region, and exited through a nozzle. Immediately before each closing of the valve to interrupt the He flow, a voltage ranging from 500 up to 1500 V was applied between the electrodes, generating an intense discharge and ejecting target electrode material by sputtering. Because of interactions with the He gas, the ejected atoms condensed into clusters that were transported through the exit nozzle by the flowing gas. Figure 3.14 shows carbon ion cluster mass spectra taken with different operational conditions. The source was able to produce cluster beams with high intensity and good stability.

3.4 Laser Irradiation Sources

Supersonic metal vapor cluster beams by laser vaporization were developed by Richard Smalley and coworkers [22]. A source they reported in 1982 is shown in Figure 3.15 [23]. This original source used a high-power pulsed laser beam focused on a small-diameter spot on a copper rod that was rotated and

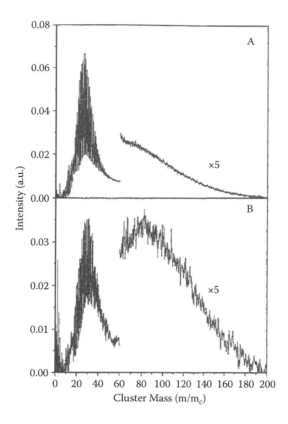

Figure 3.14 Carbon ion cluster mass spectra taken with different delays between helium injection and discharge firing.

translated continuously by a screw drive mechanism so as to avoid drilling a deep hole with the laser. The ejected copper plasma produced by the laser was entrained in helium flowing at near-sonic velocity within a narrow channel upstream from the point of free expansion into vacuum. Gas feed to this vaporization channel was through a fast solenoid-actuated pulsed valve. The work resulted in successful production of intense cold cluster beams from a wide range of elements, including Cu, Ag, Ni, Fe, Si, Mo, and W [24]. Figure 3.16 shows example TOF mass spectra of Fe, Ni, W, and Mo clusters produced by 2 mJ/cm² ArF (193 nm) excimer laser vaporization into helium at 8 atm pressure expanding through a 1 mm nozzle orifice [25].

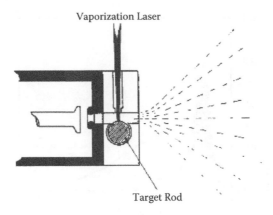

Figure 3.15 Schematic cross section of pulsed metal cluster nozzle.

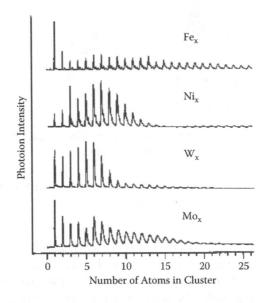

Figure 3.16 TOF mass spectra of metal clusters produced by laser vaporization with helium.

Using laser ablation of metals into high-pressure helium flowing through a nozzle cooled to cryogenic temperature has been considered one of the best ways to form cold clusters. Cold clusters are useful for materials research because a variety of precise electronic, chemical, optical, and magnetic measurements benefit greatly from removal of usual thermal effects. Compared with methods of metal cluster formation that use continuously heated vapor from ovens or crucible sources, where a major part of the condensation occurs because of cooling during a nozzle expansion, the only practical way to prepare large cold clusters is to separate the processes of vapor generation, cluster growth, and cluster cooling [26]. Michael A. Duncan has summarized 30 years of progress on laser vaporization cluster sources [27].

3.5 Ion and Plasma Sputtering Sources

Ion beams sputter atoms, molecules, and clusters from solid material surfaces. Sputtering is the result of momentum exchange between the ions and atoms in the target material during collisions. Ion beam sputtering phenomena were first observed by W. R. Grove in 1852 [28, 29]. Sputtering of materials has been widely used for thin-film deposition. Historical events during the early development have been well documented by Donald Mattox [30].

The history of cluster emission by sputtering was reviewed by Wolfgang Hofer [31]. The earliest reported observation of cluster (dimer) emission due to sputtering was recognized to have been in an experiment by Richard Honig in 1956 [32]. The clusters detected in the early experiments were usually small, comprising only a few atoms. Large clusters emitted during sputtering by accelerated Xe ion beams were reported by Itsuo Katakuse et al. in 1985 [33]. Figure 3.17a shows their sputter gun, which was intended to be used for SIMS applications [34]. The gun was operated by cold-cathode discharge and produced 5–10 µA of ion current at 10 kV. Figure 3.17b shows the distribution of Ag clusters created by direct ion bombardment. Sizes of up to 100 atoms can be seen to have been produced, and cluster sizes ranging from 50 to 250 atoms were also reported [33]. As shown in the figure, the ion intensity decreased pseudoexponentially with increasing cluster size. Katakuse also reported that two kinds of anomalies were observed to be superimposed on the pseudoexponential general decrease of the ion intensity. One was odd–even alternation, and the other was a steep decrease in the ion intensity at the magic numbers 3, 9, 21, 35, 41, 59, and 93. These

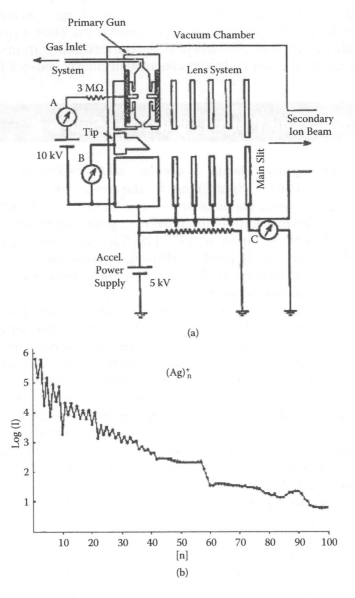

Figure 3.17 (a) Schematic diagram of the sputter ion source assembly and (b) mass distribution of silver clusters.

characteristics were explained to have resulted because the stability of clusters was strongly related to the one-electron-shell structure in which electrons were bound in a spherically symmetric potential well.

A magnetron-type metal cluster source was reported by Hellmut Haberland and coworkers in 1991 [35]. Haberland's design goal was to create a source that gave a continuous beam of variable cluster size without using a laser vaporization or a thermal evaporation process. In magnetron sputtering, magnetic fields keep the plasma in front of the target in order to intensify the ion bombardment. Inert gases, specifically argon, are usually employed as the sputtering gas. Positively charged argon ions from the plasma are accelerated toward the negatively biased target. When a positive ion collides with atoms at the surface of a solid target, energy transfer occurs. If the energy transferred to a lattice site is greater than the binding energy, a surface atom becomes sputtered. Magnetron sputtering has also been used in thin-film deposition. The principle and operation of the sources are described in several books and papers [36, 37].

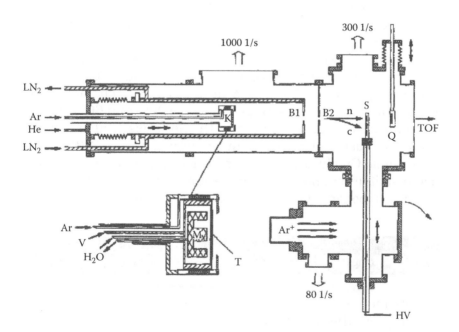

Figure 3.18 Schematic diagram of Haberland's source reported in 1994.

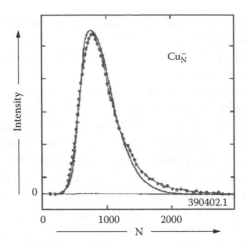

Figure 3.19 Size distribution of Cu cluster beam.

Figure 3.18 shows one of Haberland's sources as reported in 1994 [38]. The metal cluster ions were generated in a liquid-nitrogen-cooled aggregation tube by using a magnetron sputter source (K). Cluster ions generated in the tube passed through two diaphragms (B1) and (B2) and were accelerated onto the substrate (S). A Kaufman-type gun for substrate surface cleaning by Ar monomer ions was installed near the substrate area. The magnetron sputter discharge was operated at a pressure of about 100 Pa in a mixture of argon and helium. The typical discharge power was 50–200 W at about 200 V. The degree of ionization of the clusters ranged between 20% and 60%, which was unusually high compared with the levels that normally result from using electron beam ionization. Figure 3.19 shows an example Cu cluster size distribution. The maximum of the size distribution shown was at around 800 atoms/cluster; no atomic components or small clusters were detected.

3.6 Polyatomic Sources

Polyatomic ions are charged species composed of two or more of the same or different atoms. In early studies, calutron sources were used to produce polyatomic ions [39]. The calutron source was actively developed for isotope separation during the 1950s and 1960s. Using this source for polyatomic ion generation, researchers had produced gas ions of O_2, N_2, H_2, CO_2, Cl_2, and so forth [40]. Organic solvent ions of toluene, phenetole, heptane, fluorinated carbon, and other molecules [41] were produced for investigations of ion–surface interaction processes.

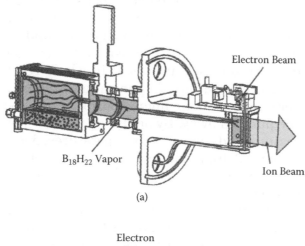

(a)

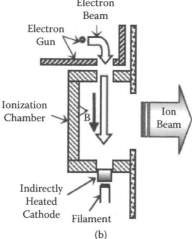

(b)

Figure 3.20 (a) Structure of cluster ion source and (b) detail of the ionization chamber.

The structure of a polyatomic cluster ion source developed for ion implantation using boron hydride compounds is shown in Figure 3.20a [42, 43]. For cluster ion beam production, when the source material was ionized by electrons supplied from the heated filament, it was required that the polyatomic gas not become dissociated by the heat from the filament. For this purpose, as shown in the detailed structure of the ion source in Figure 3.20b, the electron-supplying filament was positioned outside the ionization chamber so as to maintain a cold chamber wall. In this structure, however, feed compounds were deposited inside the ionization chamber, and the source operation time was limited. In order to prevent deposition, a self-cleaning system was used [44] in which reactive fluorine gas was introduced into the ionization chamber, and the deposits were converted to high-vapor-pressure compounds that could be pumped away. By adjusting fluorine flow, source housing pressure, and component temperatures, deposits were effectively removed from the ionization chamber, extraction electrode, and source housing, and it became possible to operate for long hours.

Figure 3.21 shows a high-resolution mass spectrum of a $B_{18}H_{22}$ ion beam. The peak at mass 220 is from 18 ^{11}B atoms plus all 22 H atoms. The rest of the peaks are composites for combinations of ^{11}B and ^{10}B atoms and different numbers of H atoms [45].

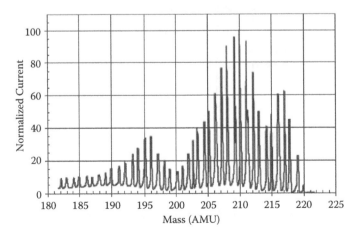

Figure 3.21 High-resolution mass spectrum of $B_{18}H_{22}$ ion beam.

References

1. P. Jena and A. W. Castleman Jr. Clusters: A bridge across the disciplines of physics and chemistry. *Proc. Natl. Acad. Sci. USA*, 103(28), 10560–10569, 2006.
2. T. H. Maugh II. When is a metal not a metal? *Science*, 219, 1413–1415, 1983.
3. F. Trager and G. zu Putlitz (eds.). *Metal Clusters*. Springer-Verlag, Berlin, 1986.
4. H. Haberland (ed.). *Clusters of Atoms and Molecules*. Springer-Verlag, Berlin, 1994.
5. J. A. Alonso. *Structure and Properties of Atomic Nanoclusters*. Imperial College Press, London, 2005.
6. K. D. Sattler (ed.). *Handbook of Nanophysics: Clusters and Fullerenes*. CRC Press, London, 2009.
7. A. H. Pfund. Bismuth black and its applications. *Rev. Sci. Instrum.*, 1, 397–399, 1930.
8. A. H. Pfund. The optical properties of metallic and crystalline powders. *J. Opt. Soc. Am.*, 23, 375–378, 1933.
9. R. Uyeda and K. Kimoto. Study of zinc black by electron diffraction (in Japanese). *Appl. Phys.*, 18, 76–78, 1949.
10. K. Kimoto, Y. Kamiya, M. Nonoyama, and R. Uyeda. An electroscope study on fine metal particles prepared by evaporation in argon gas at low pressure. *Jpn. J. Appl. Phys.*, 2(11), 702–713, 1963.
11. A. Yokozeki and G. D. Stein. A metal cluster generator for gas-phase electron diffraction and its application to bismuth, lead, and indium: Variation in microcrystal structure with size. *J. Appl. Phys.*, 49(4), 2224–2232, 1978.
12. A. Yokozeki. Lead microclusters in the vapor phase as studied by molecular beam electron diffraction: Vestige of amorphous structure. *J. Chem. Phys.*, 68, 3766–3773, 1978.
13. B. D. Boer and G. D. Stein. Production and electron diffraction studies of silver metal clusters in the gas phase. *Surf. Sci.*, 106, 84–94, 1981.
14. K. Sattler, J. Mühlback, and E. Recknagel. Generation of metal clusters containing from 2 to 500 atoms. *Phys. Rev. Lett.*, 45(10), 821–824, 1980.
15. K. Sattler. Binary metal alloy clusters. In *Metal Clusters*, eds. F. Trager and G. zu Putlitz, 123–131. Springer-Verlag, Berlin, 1986.

16. A. E. T. Kuiper, G. E. Thomas, and W. J. Schouten. Ion cluster beam deposition of silver and germanium on silicon. *J. Crystal Growth*, 51, 17–40, 1981.

17. I. M. Goldby. Dynamics of molecules and clusters at surfaces. PhD thesis, University of Cambridge, April 1996.

18. I. M. Goldby, B. von Issendorff, L. Kuipers, and R. E. Palmer. Gas condensation source for production and deposition of size-selected metal clusters. *Rev. Sci. Instrum.*, 68, 3327–3334, 1997.

19. J. Gspann. Large clusters of cesium from pure vapor expansion. Z. Phys. D., *Mol. Clusters*, 20, 421–423, 1991.

20. G. Ganteför, H. R. Siekmann, H. O. Luts, and K. H. Meiwes-Broer. Pure metal and metal-doped rare-gas clusters grown in a pulsed arc cluster ion source. *Chem. Phys. Lett.*, 165(4), 293–420, 1990.

21. E. Barborini, P. Piseri, and P. Milani. A pulsed microplasma source of high intensity supersonic carbon cluster beams. *J. Phys. D Appl. Phys.*, 32, L105–L109, 1999.

22. T. G. Dietz, M. A. Duncan, D. E. Powers, and R. E. Smalley. Laser production of supersonic metal cluster beams. *J. Chem. Phys.*, 74, 6511–6512, 1981.

23. E. Powers, S. G. Hansen, M. E. Geusic, A. C. Pulu, J. B. Hopkins, T. G. Dietz, M. A. Duncan, P. R. R. Langrldge-Smih, and R. E. Smalley. Supersonic metal cluster beams: Laser photoionization studies of Cu_2. *J. Phys. Chem.*, 86, 2556–2560, 1982.

24. D. L. Michalopoulos, M. E. Geusic, S. G. Hansen, D. E. Powers, and R. E. Smalley. The bond length of Cr_2. *J. Phys. Chem.*, 86, 3914–3916, 1982.

25. J. B. Hopkins, P. R. R. Langridge-Smith, M. D. Morse, and R. E. Smalley. Supersonic metal cluster beams of refractory metals: Spectral investigations of ultracold Mo_2. *J. Chem. Phys.*, 78, 1627–1637, 1983.

26. E. C. Honea, M. L. Homer, J. L. Person, and R. L. Whetten. Generation and photo ionization of cold Na_n clusters; n to 200. *Chem. Phys. Lett.*, 171(3), 147–154, 1990.

27. M. A. Duncan. Invited review article: Laser vaporization cluster sources. *Rev. Sci. Instrum.*, 83, 041101-1–041101-5, 2012.

28. W. R. Grove. On the electro-chemical polarity of gases. *Phil. Trans. R. Soc. Lond.*, 142, 87–101, 1852.

29. T. E. Madey. Early applications of vacuum, from Aristotle to Langmuir. *J. Vac. Sci. Technol.*, A2(2), 110–117, 1984.

30. D. M. Mattox. *The Foundations of Vacuum Coating Technology*. Noyes/Williams Andrew Publishing, Norwich, NY, 2003.

31. W. O. Hofer. Angular, energy, and mass distribution of sputtered particles. In *Sputtering by Particle Bombardment III Characteristics of Sputtered Particles: Technical Applications*, eds. R. Behrisch and K. Wittmaack, 15–90. Springer-Verlag, Berlin, 1991.

32. R. E. Honig. Sputtering of surface by positive ion beams of low energy. *J. Appl. Phys.*, 29(3), 549–555, 1956.

33. I. Katakuse, T. Ichihara, Y. Fujita, T. Matsuo, T. Sakurai, and H. Matsuda. Mass distributions of copper, silver and gold clusters and electronic shell structure. *Int. J. Mass Spectrom. Ion Process.*, 67, 229–236, 1985.

34. I. Katakuse, T. Ichihara, H. Nakabush, T. Matsuo, and H. Matsuda. A compact primary gun for molecular SIMS. *Mass. Spectrosc.*, 31(2), 111–114, 1983.

35. H. Haberland, M. Karrais, and M. Mall. A new type of cluster and cluster ion source. *Z. Phys. D*, 20, 413–415, 1991.

36. J. L. Vossen and W. Kern (eds.). *Thin Film Processes*. Academic Press, Boston, 1978.

37. S. M. Rossnagel, J. J. Cuomo, and W. D. Westwood (eds.). *Handbook of Plasma Processing Technology*. Noyes Publications, Park Ridge, NJ, 1990.

38. H. Haberland, M. Mall, M. Moseler, Y. Qiang, T. Reiners, and Y. Thurner. Filing of micron-sized contact holes with copper by energetic cluster impact. *J. Vac. Sci. Technol. A*, 12(5), 2925–2930, 1994.

39. A. Theodore Forrester. *Large Ion Beams*. John Wiley & Sons, Hoboken, NJ, 1988.

40. S. R. Kasi, H. Kang, C. S. Sass, and J. W. Rabalais. Inelastic processes in low-energy ion-surface collisions. *Surface Sci. Rep.*, 10, 1–104, 1989.

41. R. G. Cooks, T. Ast, and MD. A. Mabud. Collisions of polyatomic ions with surfaces. *Int. J. Mass Spectrom. Ion Process.*, 100, 209–265, 1990.

42. T. N. Hrosky, Universal Ion Source™ for cluster and monomer implantation. In *2006 International Conference on Ion Implantation Technology Proceedings*, eds. K. J. Kirkby, R. Gwilliam, A. Smith, and D. Chivers, 159–162. American Institute of Physics, New York, 2006.

43. A. Renau, A. Perel, M. E. Mack, and T. N. Horsky. Ion sources. In *Ion Implantation Technology, Science and Technology*, ed. J. Ziegler, 6-1–6-54. Ion Implantation Technology Co., Chester, Maryland, USA, 2010.

44. T. N. Horsky, G. F. R. Gilchrist, and R. W. Milgate III. Boron beam performance and in-situ cleaning of the ClusterIon® source. In *2006 International Conference on Ion Implantation Technology Proceedings*, eds. K. J. Kirkby, R. Gwilliam, A. Smith, and D. Chivers, 198–201. American Institute of Physics, New York, 2006.
45. D. Jacobson, T. Horsky, W. Krull, and B. Milgate. Ultra-high resolution mass spectroscopy of boron cluster ions. *Nucl. Instrum. Methods Phys. Res. B*, 237, 406–410, 2005.

4

Gas Cluster Ion Beam Equipment

Technological development of gas cluster ion beam (GCIB) equipment has required innovations in a number of areas, including in generation of cluster beams, ionization of clusters, and cluster ion transport. In this chapter, the development work that has been conducted on these topics is described, and results of fundamental studies that contributed to the necessary understanding of the characteristics of clusters are discussed.

4.1 Gas Cluster Ion Beam Equipment for Process Applications

In order to produce a beam of energetic gas cluster ions, GCIB equipment employs a gas expansion nozzle to generate neutral clusters, a skimmer to transmit a directed axial stream of neutral clusters while blocking flow of excess gas from the nozzle, an assembly to ionize the clusters, and electrodes at high potentials to accelerate the ionized clusters to high energies. Because the gas flows used for cluster generation are relatively high, typically several hundred standard cubic centimeters per minute (sccm) for most gases, separate pumps are usually employed for the nozzle, ionization/acceleration, and target chamber stages of a GCIB system. Sometimes, a differential

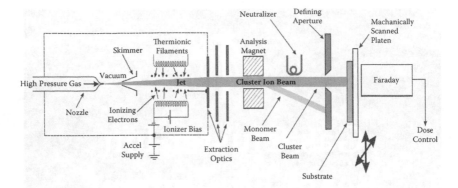

Figure 4.1 Schematic representation of a typical GCIB processing system.

pumping stage is used between the cluster nozzle source and the ionization chamber in order to allow high gas flows through the cluster generation nozzle while still maintaining acceptably good vacuum in the cluster ionization and acceleration regions. Additional lens assemblies can be utilized to improve the efficiency of cluster ion transport to target substrates. Electrostatic or magnetic filtering is employed to eliminate monomer ions and small cluster ions that behave differently than large cluster ions upon target impact. Uniform processing of large-area substrates is accomplished by means of electrostatic scanning of the cluster ion beams or by mechanical scanning of substrates within a fixed beam. Faraday current monitors allow accurate cluster ion dose control. Low-energy electron flood sources can be used to prevent charge buildup on surfaces undergoing processing. Figure 4.1 shows a schematic representation of a typical GCIB processing system [1–3].

4.2 Generation of Gas Cluster Beams

Gas cluster formation involves a gas expanding from a high-pressure source into vacuum through a small orifice (see Section 2.1). Clusters from many gases and compounds having high vapor pressures can be generated by using supersonic nozzles. Supersonic expansion can produce high-intensity beams containing large numbers of clusters. De Laval nozzles

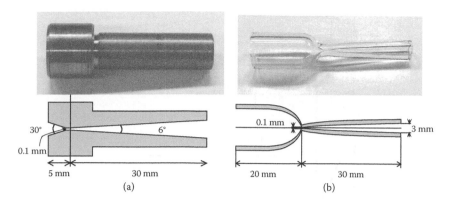

Figure 4.2 Typical nozzle sources: (a) conical nozzle and (b) de Laval nozzle.

and converging–diverging nozzles are considered to be efficient because they have contours designed for efficient cluster nucleation through control of the rate of gas expansion.

Parameters related to nozzle shape, including throat diameter, divergence angle, and length, are important in determining beam intensities and cluster size distributions. Figure 4.2 shows examples of representative nozzles that have been used with many different gases to produce clusters of sizes ranging from small to very large with many thousands of atoms. In all these nozzles, condensation of vapor is essential for cluster formation.

In the cluster generation, two types of condensation and nuclei formation are important: homogeneous nucleation and heterogeneous nucleation. Homogeneous nucleation takes place when the gaseous vapor is free of all impurities such that condensation cannot occur at the saturation point and the vapor can exist in a state of supersaturation. Because vapor atoms or molecules in the supersaturation region still move in a random manner and collide with each other, collisions cause aggregates, embryos, nuclei, and droplets to form in the system. Heterogeneous nucleation occurs when vapor reaches saturation pressure and temperature, and vapor can then condense at its saturation point in the presence of sufficient amounts of available foreign condensation nuclei, such as particles, ions, and surfaces [4].

Figure 4.3 shows an example of static pressure variation (P) along the axis of a supersonic nozzle (P_0 denotes stagnation

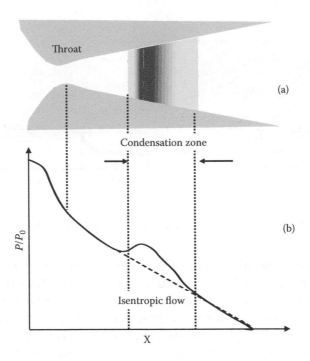

Figure 4.3 Typical static pressure variation along the axis of a super-sonic nozzle.

pressure). The presence of homogeneous condensation causes a change of gas pressure and gas temperature within the nozzle. The onset of condensation and cluster formation can be recognized to occur in the region where pressure deviates from the expansion line (as was seen in Yellot's early experiment and was shown in Figure 2.3). At the point of abrupt pressure rise, a very high supersaturation condition is created, and growth of nuclei into clusters becomes dominant.

In theoretical analyses, pressure and temperature distributions along nozzle axes can be calculated by first using one-dimensional gas dynamic flow equations, assuming conservation of mass, momentum, and energy, and then including nucleation rate and droplet formation equations [5–7]. The pressure (P) distribution and temperature distribution, as well as the Mach number, can be estimated at every point along a nozzle axis by using the stagnation temperature, stagnation pressure (P_o) of the gases, and the geometry of the nozzle.

Such a simple analysis gives a basic idea of the supersonic expansion phenomena and cluster formation mechanisms in the supersonic nozzle. Cluster beams can be generated from many gases and compounds having high vapor pressures by using supersonic nozzles that exploit the adiabatic expansion characteristic of a gas expanding into vacuum through a small aperture. The supersonic expansion approach has been most successful in producing high-intensity beams containing large numbers of clusters.

Figure 4.4 compares calculated flow distributions through two nozzles: a de Laval nozzle and a conical nozzle. The calculations were performed using a Direct Simulation Monte Carlo (DSMC) method [8]. Lines on these two plots are streamlines reflecting the flow intensity and flow direction. Figure 4.4b shows that an intense directional beam forms in the conical

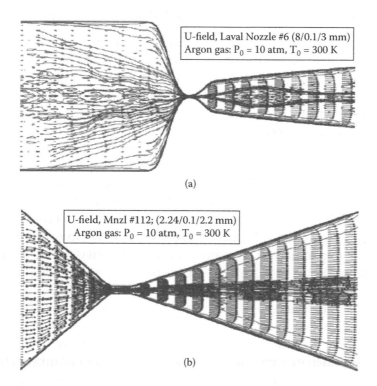

U-field, Laval Nozzle #6 (8/0.1/3 mm)
Argon gas: P_0 = 10 atm, T_0 = 300 K

(a)

U-field, Mnzl #112; (2.24/0.1/2.2 mm)
Argon gas: P_0 = 10 atm, T_0 = 300 K

(b)

Figure 4.4 Comparison between two nozzle shapes: (a) de Laval nozzle and (b) conical nozzle.

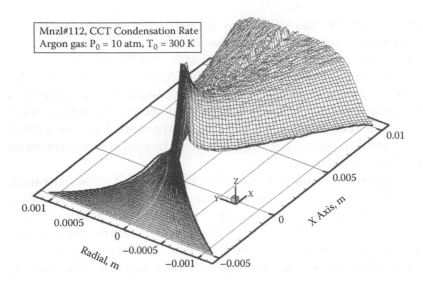

Figure 4.5 Condensation rate for the conical nozzle from Figure 4.4b.

nozzle. Figure 4.5 shows the condensation rate in the conical nozzle shown in Figure 4.4b as calculated using the DSMC analysis combined with classical condensation theory [9]. It becomes clear that condensation of gases occurs most strongly near the nozzle throat and then gradually decreases through the conical expansion region.

After expansion of the gas out of a nozzle, a skimmer having the shape of a truncated cone is usually employed to separate the cluster-containing core of the gas stream from the waste gas. Skimmers typically have entrance aperture edges made as sharp as possible to avoid collisional distortion of the impinging gas flows. Figure 4.6 shows an experimental result of the dependence of the neutral beam intensity on the source gas stagnation pressure at temperatures T_0 = 300K and 220K for an example de Laval nozzle having a throat diameter of 0.1 mm and a length of 30 mm. In the case shown, the flow rate of Ar was 600 sccm at a source gas pressure of 4000 Torr. The neutral beam intensity can be determined by observing the difference in vacuum pressure at the target chamber when the beam shutter is opened or closed. The neutral beam intensities increased monotonically starting from a P_0 of 1000 Torr. When the source gas temperature was lowered from 300K to

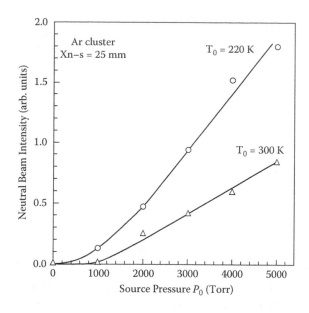

Figure 4.6 Source pressure and temperature dependence of neutral beam intensity.

220K, the neutral beam intensity at the same P_0 was twice as high as that at 300K. High gas pressures and low gas temperatures are important for obtaining intense cluster beams.

Researchers have studied the formation of clusters from a variety of gases that are available for surface modification. From Ar and CO_2 gases, cluster beams are usually produced at room temperature using conical or de Laval–type nozzles. From other gases such as N_2, O_2, and SF_6, directly forming intense cluster beams at room temperature is often difficult. In order to produce strong cluster beams from such gases, instead of employing expansion of the pure gas, He gas can be mixed with the desired gas as a carrier. The role of the He, which does not become incorporated into the clusters, is to assist in heat removal during the nozzle expansion. Figure 4.7 shows cluster beam intensities from a room temperature nozzle for SF_6, CO_2, Ar, N_2O, N_2, and O_2 gases mixed with He gas as functions of the gas ratios. It has been observed that neutral cluster beam intensities can be increased by up to two orders of magnitude by incorporating a modest amount of He into the source gas. For semiconductor device applications, such as doping and

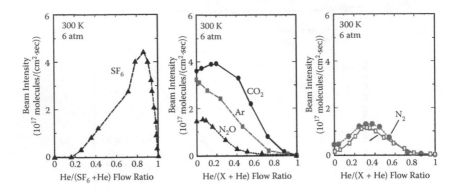

Figure 4.7 Cluster beam intensities as functions of He gas mixture ratios.

alloying, gas mixtures with a small percentage (1–5%) of a doping gas (e.g., B_2H_6) or a mixture of gases (e.g., $GeH_4 + B_2H_6$) diluted in an inert gas such as Ar or He have been used [10]. By proper choice of the gas used for cluster generation and by proper selection of acceleration voltage and cluster ion dose, many processes are possible by GCIB. Examples of processes and suitable gases include the following:

- Smoothing or asperity removal: Ar, O_2, N_2.
- Oxide, nitride, or carbide formation: O_2, N_2, NH_3, N_2O, CO_2, CH_4.
- Etching or ashing: NF_3, O_2, SF_6.
- Doping and alloying: B_2H_6, BF_3, PH_3, AsH_3, GeH_4, SiH_4.

4.3 Ionization of Gas Cluster Beams

Ionization of neutral gas clusters is usually performed by electron bombardment. Electrons emitted from tungsten filaments are accelerated by a voltage applied between the filaments and a cylindrical anode so that they collide with gas clusters passing within the interior of the anode, causing ejection of electrons from some of the clusters and making those clusters positively charged. A schematic illustration of a typical ionizer is shown in Figure 4.8 [11]. Other configurations of ionizers have been developed that incorporate a self-neutralizing effect to avoid

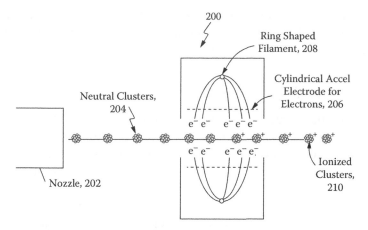

Figure 4.8 Schematic illustration of typical gas cluster ionizer.

space charge blowup of the cluster ion stream within the ionizer [12, 13]. In these designs electrons extracted from the filaments pass through the cluster jet and then strike an opposing electrode to produce low-energy secondary electrons that help ensure that the ionized jet remains space charge neutral.

Gas cluster beams produced by supersonic expansion usually have broad cluster size distributions with hundreds or thousands of atoms per cluster. GCIB cluster size measurements are usually done by time-of-flight (TOF) methods using electrostatic chopping of the beams. Several other methods for size distribution characterization are also available, such as TOF using a mechanically chopped rotating disk with a small hole, a retarding field method, a method employing observation by Debye–Sherrer electron diffraction, and a gas scattering method [14].

The size distributions measured by TOF for clusters ionized by electron impact reflect not only the distributions of the neutral precursors as formed by supersonic expansion, but also factors such as size-dependent ionization cross sections, ion stabilities, and detection efficiencies [15]. The effects of electron impact have been studied in terms of ionization mechanisms, ionization efficiency, and multiple ionization probabilities, and also in terms of stability, decay processes, fragmentation phenomena, and evaporation behavior. Several review papers have described these characteristic behaviors in detail for various cluster ions [16–21]. For engineering applications, cluster size distributions and their charge state conditions are of great importance.

4.3.1 Electron Energy Dependence

Electron impact energy is an important factor for ionization of cluster beams with large cluster sizes. Cluster beams produced by supersonic expansion contain broad distributions of different masses, cluster size N, and the clusters can be ionized by electron impact to become singly charged or multiply charged ions, charge state z. In 1973, the effects of electron energy in formation of multiply charged cluster ions were observed by J. Gspann and K. Körting [22]. Their cluster beams were formed by expansion of gas that was precooled to the liquid phase. The beams were directed through a conical skimmer and collimators and then were chopped by a rotating disk. The TOF signals were detected by a conventional mass spectrometer with electron impact ionization. Figure 4.9 shows TOF distributions of accelerated nitrogen cluster ions for different ionizing electron energies. The smooth curves were obtained by convolution of the measured signals. Vertical lines denote arrival times of singly and multiply charged ions. Increasing the electron energy to beyond about 30 eV caused new peaks to appear at earlier arrival times corresponding to lower mass-to-charge ratios. The mean size of singly charged clusters produced by using 34 eV electrons was approximately 1.2×10^4 atoms/cluster. At higher electron energies, up to 10 distinct mass-to-charge peaks could be identified.

M. E. Mack et al. observed the effects of ionization conditions on cluster ion size distributions using TOF measurements [23]. At low electron energies and currents, where singly charged clusters would be expected, the distribution peaks were observed at 15,000 or more atoms/charge. At higher energies and currents, however, the peaks shifted down to 2000 to 5000 atoms/charge. Similar examples of the effects of ionization conditions on the cluster ion size distributions have been reported by T. Seki et al. [24]; at higher electron ionization energies, they observed that the cluster ion mass to charge size distributions decreased. These results confirm that cluster ion

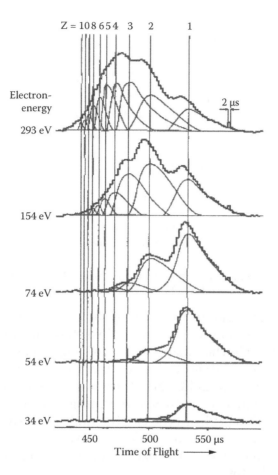

Figure 4.9 TOF distributions of accelerated cluster ions and charge state z for different ionization electron energies.

beams containing large size clusters will normally contain considerable numbers of multiply ionized clusters.

Earlier experimental investigations of the effects of electron impact energy on the formation of singly and multiply charged clusters have been reported [25, 26]. Figure 4.10 shows currents of singly, doubly, and triply charged Ar monomers and size-selected Ar clusters versus ionizing electron energies at values close to the ionization thresholds [18]. Threshold electron energies for creation of doubly charged Ar_{101}^{2+} and triply charged Ar_{236}^{3+} cluster ions were found to be substantially lower than the onset energies for the respective Ar^{2+} and Ar^{3+}

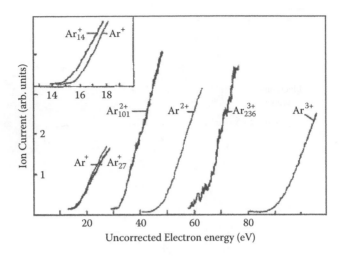

Figure 4.10 Ion current versus electron energy close to threshold for the production of Ar atom and Ar cluster ions.

monomer ions. Also, in the case of the cluster ions, the onset energies required for double and triple charging were roughly double and triple the onset energy for single ionization. This finding was understood to mean that the production of doubly ionized Ar_n^{2+} should be attributed to two sequential single ionization events caused by one incoming electron hitting two different Ar atoms within the cluster, and similarly in the case of Ar_n^{3+}, three sequential single ionization events caused by the same incoming electron exciting three different Ar atoms within the cluster.

4.3.2 Cluster Size Dependence

It is believed that critical sizes exist for multiply charged cluster ions. If small clusters become multiply ionized, electric repulsion forces between the charges in the cluster can exceed the binding forces that keep the cluster atoms together and the cluster will immediately fragment. This is called Coulomb explosion, and multiply charged cluster ions smaller than critical size do not survive. The first evidence for Coulomb explosion of microclusters was reported by K. Sattler and coworkers in 1981 [25]. Their paper stated the following: "The two positive charges generated at one atom by electron bombardment are likely to move to opposite sides of the cluster which then explodes into singly charged fragments as long as the Coulomb repulsion energy is greater than the binding energy. Only if the particle size exceeds a critical value are doubly charged clusters stable."

P. Scheier and T. D. Märk used a high-resolution mass spectrometer to investigate the balance between Coulomb repulsion and surface energy as a function of cluster size for relatively small Ar clusters [17]. As shown in Figure 4.11, their spectrometer mass spectra from charged Ar clusters showed intermediate peaks between Ar_{46}^+ and Ar_{47}^+ and for all larger masses. These peaks were unquestionably identified as doubly charged Ar_n^{2+}. The critical size deduced in this experiment for the occurrence of doubly charged argon cluster ions in the direct mass spectrum was $N = 91$. For triply charged cluster ion Ar_n^{3+}, the critical size was $N = 226$, determined by the same detection method [18].

To estimate critical sizes where multiply charged clusters start to be produced, D. Kreisle et al. developed a model based on a liquid-drop approximation [26]. Their calculated results

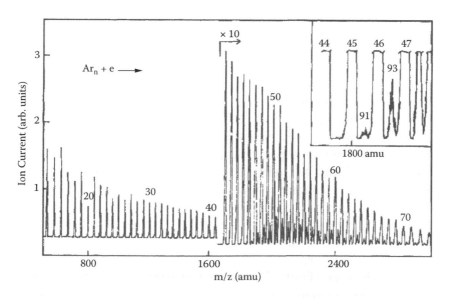

Figure 4.11 Mass spectrum of Ar cluster beam ionized with 80 eV electron energy.

TABLE 4.1
Calculated Critical Sizes, N_c (z), of Doubly ($z = 2$),
Triply ($z = 3$), and Quadruply ($z = 4$) Charged
van der Waals Clusters

Compound	Nc (2)	Nc (3)	Nc (4)
He	17092	76895	198966
Ne	868	2950	6424
Ar	122 (91)	333 (226)	648
Kr	71 (71)	177 (156)	331
Xe	46 (51)	107 (114)	196 (208)
H_2	863	2824	6014
N_2	145 (99)	411 (215)	814
O_2	120 (92)	330	646
CO_2	54 (45)	143 (109)	274 (216)
N_2O	44 (51)	104 (105)	191 (184)
SF_6	44 (39)	121	234

Source: Depicted from O. Echt et al., *Phys. Rev. A*, 38(7),
       3243, 1988, Table II of reference [19] (modified).
Note: Experimental values are shown in parentheses.

were reported for Ar, Kr, Xe, and CO_2 clusters. O. Echt et al. subsequently reported additional liquid-drop model results for various other van der Waals clusters [19, 27]. Table 4.1 lists calculated critical sizes of doubly, triply, and quadruply charged clusters for a few of the many gases that are included in Table II of [19]. For Ar clusters, calculated critical sizes for Ar_n^{2+}, Ar_n^{3+}, and Ar_n^{4+} were 122, 333, and 648 atoms, respectively. In Table 4.1, experimentally measured critical size values are also shown in parentheses for some of the gases. The results calculated by using the liquid-drop model show reasonably good agreement with the experimentally measured critical sizes in most cases. However, the investigators pointed out that for very weakly bound clusters (Ar, N_2, O_2, and CO_2), their model tended to overestimate the critical sizes of doubly charged clusters, and the agreement was worse for triply charged clusters. They also predicted that the fission of quadruply charged clusters might yield either two doubly charged ions or one singly charged ion plus a triply charged ion. The model correctly presented many of the essential features of the ionization and fission processes, even though simplifying assumptions based on easily obtainable bulk properties had been used.

4.3.3 Ionization Efficiency Dependence

Clusters are ionized by electron impact. Ionization efficiency is related to the electron impact cross sections. Bottiglioni et al. studied ionization cross sections of H_2, N_2, and CO_2 clusters [20]. They reported that the cross sections for ionization by electron impact increase in proportion to the cluster size N for small clusters of size up to roughly $N = 50$; above this number, the cross sections increase approximately in proportion to $N^{2/3}$.

Ionization efficiency of GCIB equipment has been estimated by using a TOF spectrometer technique [28]. Figure 4.12 shows a schematic diagram of the apparatus that was constructed to integrate a TOF spectrometer (B) at the exit of the beamline of a GCIB system (A). The GCIB source was followed by an electrostatic deflector designed to remove all ionized clusters from the beam entering into the TOF chamber so that the spectrometer was able to observe the mass distribution spectra of only the neutral clusters. Efficiency of ionization as a function of cluster size was then determined by comparing the mass distribution spectra obtained without and with the GCIB ionizer operating.

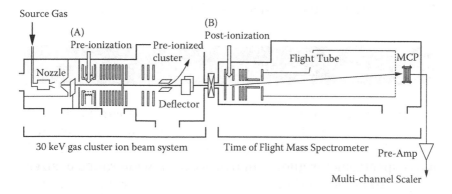

Figure 4.12 Schematic diagram of the TOF mass spectrometer connected to the GCIB equipment.

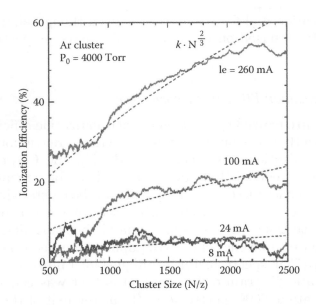

Figure 4.13 Cluster size dependence of the ionization efficiencies.

Figure 4.13 shows experimental measurements of Ar cluster ionization efficiency as a function of cluster size at a fixed ionization energy of 300 eV for several values of ionization electron current (I_e). Dotted lines shown in the figure represent fitting functions $kN^{2/3}$, where k is an arbitrary constant and N is the cluster size. The ionization efficiency was found to increase with cluster size regardless of the ionization electron current. As the cross section of a cluster with size N is proportional to $N^{2/3}$, a cluster is much easier to ionize than is a monomer.

4.4 Cluster Size Distributions

Clusters generated by the methods described can have large sizes of up to 10,000 atoms or more. In order to characterize cluster size distributions, simple equipment such as a TOF mass spectrometer allows analyses over a wide range of sizes. A spectrometer method can provide high-resolution spectra for small clusters. For very broad cluster distributions and very large cluster sizes, resolutions are lower but the method remains as a convenient way to obtain a qualitative view of the distributions.

A typical TOF mass spectrum of Ar clusters with sizes of up to 35 atoms is shown in Figure 4.14. The source pressure in this situation was 4000 Torr, and the ionization energy was 70 eV. The spectrum shows many peaks at intervals of 40 atomic mass units, which is the mass of an Ar atom. The intensity of Ar_n can be seen to decrease exponentially with cluster size. Figure 4.15 shows TOF wide-range mass spectra taken at different source gas pressures P_0 from 760 to 3800 Torr. The ionization energy and the acceleration energy for all spectra were 70 and 1.5 keV, respectively. At P_0 of 760 Torr, the cluster beam intensity can be seen to decrease suddenly with cluster size, starting from monomers. In this condition, the clusters grow by monomer additions. At P_0 of 1500 Torr, the decay is no longer exponential, and there is a small maximum at a size of around 500 atoms/cluster.

With increasing P_0, the position of the peak of the size distribution can be seen to shift to larger sizes, and the dip observed at the size of about 200 atoms becomes more pronounced. The mass spectrum is expressed as the combination of two distributions, exponential decay in the small size region and wide distribution in the larger cluster region. When P_0 is

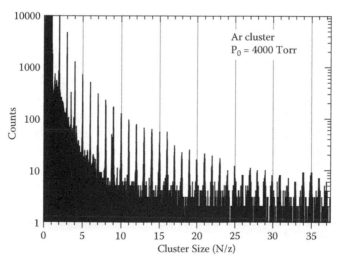

Figure 4.14 TOF mass spectrum of Ar clusters with sizes of up to 35 atoms.

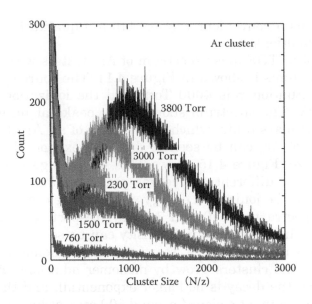

Figure 4.15 TOF spectra of Ar large clusters at different source gas pressures.

high, clusters grow not only by monomer additions, but also by cluster–cluster aggregation, which contributes to the broad distribution at larger sizes. Beyond a critical size N^*, cluster–cluster aggregation becomes the dominant mechanism for growth of large clusters. The critical size N^* can be estimated from the dip in the TOF spectra to be around 200 atoms/cluster at P_0 above 1500 Torr. As can also be observed in Figure 4.15, both the total number of clusters and their sizes increase significantly with P_0 above 1500 Torr. Clusters of the sizes produced by room temperature nozzles were found to be particularly useful for materials processing because the energy per atom is a few electron volts to a few tens of electron volts. This is the favorable energy range for nanotechnology materials processing applications. Recent experiments conducted with size-selected cluster beams have confirmed the fortuitous nature of the cluster size distributions that are typically produced.

4.5 Beam Transport Characteristics of Cluster Ion Beams

The kinetic energy, cluster size, and charge state are the most important characteristics of GCIB. GCIB produces low-energy bombardment effects since the kinetic energy of each atom in a cluster ion is roughly equal to the energy of the cluster divided by the number of atoms contained in the cluster. This characteristic allows use of high-voltage beam extraction and subsequent transport of the beam at high energy to the target.

According to the Child–Langmuir law, the maximum current density J_{max} that can be extracted from a planar diode is related to the extraction voltage (V) and particle mass (m) and is limited by $J_{max} \sim V^{3/2}/m^{1/2}$ [29]. Extraction of ions with energies of only a few hundreds of electron volts from a source is very difficult. Cluster ion beams can solve this problem.

In regard to cluster size, a cluster ion of size N transports N times as many atoms as does a single-monomer ion. At the same energy per atom, however, the cluster ion can be extracted at N times higher extraction voltage. Therefore, the relative maximum atom flux with GCIB can be expressed as the following [30]:

$$\text{Atom flux} \sim N \times (NV)^{3/2}/(Nm)^{1/2} = N^2 \times J_1$$

where N is the cluster size and J_1 is the current density of atomic ions that have the same energy per atom as the cluster, and hence penetrate to about the same depth.

The number of atoms delivered to the target can thus be increased by a factor of N^2 by employing clusters of N particles instead of monomers of the same species. A similar calculation can be done to compare B and $B_{10}H_{14}$ implantation. For the same B penetration, the extraction voltage of a $B_{10}H_{14}$ ion must be higher by the ratio of the ion masses; therefore, the beam current can be 122.2/10.8 = 11.3 times higher than that of a

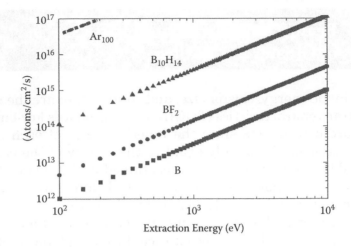

Figure 4.16 Child–Langmuir limit atom flux from an ion source versus extraction energy for B, BF_2, $B_{10}H_{14}$, and Ar $_{100}$ ions.

B monomer ion beam. The actual B flux to the surface will be another 10 times higher than this, as each cluster contains 10 B atoms. The B delivery to the surface can then be as much as $10 \times 11.3 = 113$ times higher in the case of $B_{10}H_{14}$. The maximum B flux density (or dose rate) for B, BF_2, $B_{10}H_{14}$, and Ar_{100} is plotted in Figure 4.16.

The space charge balance in a high-current beam is a well-known difficult problem in the transport of low-energy ions. Without proper beam neutralization, coulombic repulsion among ions in the beam will lead to beam expansion and lower beam current on the target. The effect becomes increasingly severe as the beam energy decreases or the beam current increases. Cluster beams can provide higher mass transport capability before being excessively affected by space charge effects and beam blowup. A cluster ion beam is less susceptible to these problems because the ions are at higher energy and the mass flux at a given beam current is very high, as determined by the average number of atoms in each cluster. Reduced beam expansion means more uniform implantation, lower ion loss on walls or electrodes, and simplified beam neutralization and focusing. A further advantage of cluster ions can be a reduction in surface charging problems. The ion charge arriving at a target surface per atom of arriving material is reduced as

TABLE 4.2
Classification of Clusters according to Bonding Type

Type	Example	Binding Type	Average Binding Energy
Van der Waals	Rare gases R_n, $(N_2)_n$, $(CO_2)_n$, $(SF_6)_n$	Weak intermolecular Van der Waals bond	≤0.3 eV
Molecular	Organics $(M)_n$, $(I_2)_n$	Weak disperse electrostatic	~0.3–1 eV
H-bonded	$(HF)_n$, $(H_2O)_n$	H-bonding electrostatic	~0.3–0.5 eV
Ionic	$(NaCl)_n$, $(CaF_2)_n$	Ionic bonds	~2–4 eV
Valence	C_n, S_8, As_4	Covalent chemical bonds	~1–4 eV
Metallic	$(W)_n$, $(Ag)_n$, $(Na)_n$, $(Al)_n$, $(Sb)_n$, $(Cu)_n$	Metallic bonds	~0.5–3 eV

Source: Modified from J. Jortner, *Phys. Chem.*, 88, 188, 1984, Table 1.

the size of the cluster increases, by three orders of magnitude for the case of singly charged clusters with $N \sim 1000$. This can greatly simplify the task of controlling surface charging on wafers by means of electron or plasma flooding.

The binding energy of atoms in a cluster has a strong influence on the changes of energy, size, and charge during transportation. Clusters of different substances can have different bonding types depending on their size. Table 4.2 shows the classification of clusters according to their bonding types of several substances [31]. Ar clusters consist of Ar atoms weakly bonded by van der Waals forces. Van der Waals bonding occurs when fluctuations in the electron density of the atom in the cluster cause instantaneous electronic dipoles, which create dipoles in nearby atoms. Rare gases, such as Ar and Xe, $(N_2)_n$, $(CO_2)_n$, and $(SF_6)_n$, form van der Waals clusters, and the average binding energy is usually lower than 0.3 eV/atom. Binding energies of molecular clusters such as hydrogen-bonded clusters $(HF)_n$ and $(H_2O)_n$ are higher than those of van der Waals clusters, and their binding energies are usually in a range of 0.3–1 eV/atom.

Collisions between GCIB ions and residual gas have to be considered seriously for cluster beam transport. While it is important to increase the flow rate of source gas and form a strong neutral beam to obtain high current GCIB, the pressures in the ionization and process chambers have to be kept at high vacuum. For this purpose, pumps with a high pumping speed have to be installed in these vacuum chambers. One of the reasons for such a requirement of a good vacuum is that the collision frequency of clusters is high because of their large cross sections. Another reason is that Ar atoms in an Ar cluster ion are weakly bonded with van der Waals forces. If the vacuum in the beam line is poor, Ar cluster ions will frequently collide with residual gas molecules and will collapse.

In order to study the effects of GCIB collision with gas, measurements of the energy and the velocity of GCIB have been made using a size-selected GCIB system combined with a charge, energy, and mass analyzer (QEM analyzer) [32, 33]. The cluster size selection is performed by magnetic separation by means of a NdFeB permanent magnet that achieves 1.2 T on its center axis. Figure 4.17 shows a schematic diagram of the system [34]. The magnet has a 50 mm gap with a 6.6° wedge angle to give vertical focusing. The effective field length is 450 mm. Gas cluster ions are bent by this magnetic field, and the atomic mass of the cluster ions can be determined from the acceleration voltage and the focus position. In order to evaluate the performance of the magnet, a TOF mass spectrometer was mounted on the scanning rail at the position

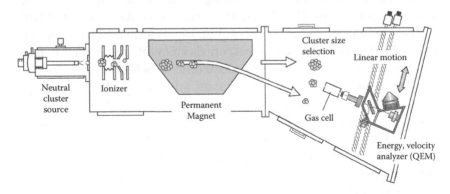

Figure 4.17 Experimental setup of size-controlled GCIB system with QEM and gas cell.

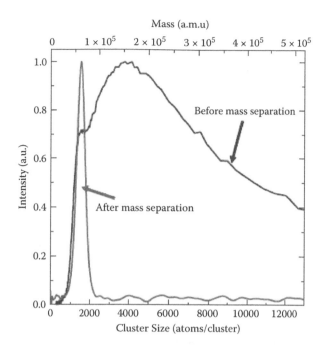

Figure 4.18 TOF spectra of Ar-GCIB before and after mass separation.

where the QEM was shown in Figure 4.17. The measurement was made at the position corresponding to the cluster size of 1500 atoms/cluster. Figure 4.18 shows the TOF mass spectrum of the 5 keV Ar cluster ion beams. The result showed that the cluster size distribution with the magnet filter was approximately 1/20 of the mean size of the original cluster distribution, and it was confirmed that the size-selected GCIB would be sufficient for further experiments concerning the collision of GCIB with residual gas.

By positioning the QEM analyzer and the gas cell at the position where the desired cluster size was obtained, cluster size dependence on energy loss and velocity after the collisions with the residual gas was studied. A collision cell (gas cell) is installed in front of the QEM analyzer. The length of the gas cell is 140 mm, and it has apertures 5 mm in diameter at both ends. Ar gas can be introduced directly from outside of the chamber controlled with a variable leak valve. Details of the QEM are shown in Figure 4.19. Even if the pressure inside of the gas cell is increased to 1×10^{-4} Torr, there is no

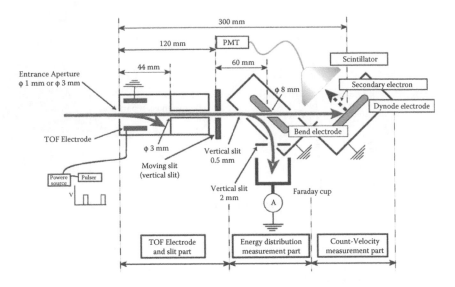

Figure 4.19 Schematic diagram of QEM apparatus with particle trajectories.

change of the vacuum pressures in the ionization and process chambers, which remain at 2×10^{-6} Torr or lower.

The total acceleration voltage could be varied from 10 to 30 kV. The ionizer was operated under conditions where the charge on the Ar cluster ions was close to 1. The energy distributions and the velocity were measured by an electrostatic energy analyzer and by a TOF method.

Figure 4.20 shows the peak energy change as a function of vacuum pressure (P_g) in the gas cell. The Ar cluster size (N/z) was 1000. Cluster ions that were accelerated at a high voltage of 30 kV lost kinetic energy with increasing P_g faster than those accelerated at 10 kV. In the case of the acceleration voltage of 10 kV, there was a 45% decrease of energy from 9.0 to 5.0 keV as P_g was increased from 1.6×10^{-6} to 1×10^{-4} Torr. In the case of 30 kV acceleration, the energy of GCIB decreased 73% from 24.2 to 6.5 keV due to the same increase of P_g. This result indicated that a cluster that has high velocity loses energy more easily than does an identical cluster having lower velocity. However, there was no substantial change in the velocity that was measured by TOF for all of the acceleration voltages, even though the pressure in the gas cell was

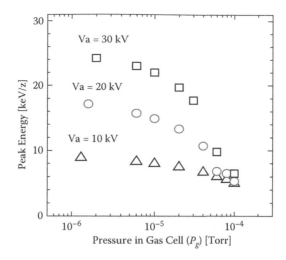

Figure 4.20 Peak energy change of Ar-GCIB as a function of the vacuum pressures in a gas cell (P_g).

varied. This suggests that the center of mass of the cluster continues to travel at the same velocity after collisions with residual gas. Thus, energy loss of a cluster is caused by releasing of constituent atoms and, as a result, decreasing of cluster size.

The energy loss is also affected by the gas species or bonding strength of clusters. Figure 4.21 compares the energy changes between Ar and CO_2 cluster ion beams for various cluster sizes. It can be seen that as background gas pressure increased, the larger Ar or CO_2 clusters did not lose energy as rapidly as did the smaller clusters of the same gas. Also, Ar clusters lost more energy with increasing pressure than did CO_2 clusters of the same size. Since the binding energy of solid CO_2 is stronger than that of solid Ar, the CO_2 cluster ion does not collapse as easily as a result of the collisions with residual gas. When the bonding of atoms or molecules in a cluster is strong, such as in the case of CO_2, the cluster ion keeps its kinetic energy despite frequent collisions with residual gas. Because of the large collision cross section and the weak bonding strength between atoms in a cluster, the operational conditions of GCIB equipment, such as nozzle gas pressure, ionization electron energy, acceleration voltage, and

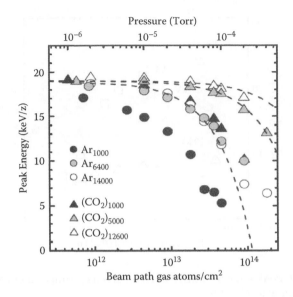

Figure 4.21 Peak energy change of Ar and CO_2-GCIB as a function of P_g.

gas pressure in the vacuum chamber, are all important. Many unique aspects of cluster ion beam formation and transport remain to be discovered.

References

1. W. Skinner. Status of GCIB for manufacturing applications. In *Extended Abstracts of the Workshop on Cluster Ion Beam Process Technology*, Tokyo, October 15–16, 2001, pp. 23–24.
2. A. Kirkpatrick. Gas cluster ion beam applications and equipment. *Nucl. Instrum. Methods B*, 206, 830–837, 2003.
3. I. Yamada, J. Matsuo, N. Toyoda, and A. Kirkpatrick. Materials processing by gas cluster ion beams. *Mater. Sci. Eng.*, R34, 231–295, 2001.
4. S. S. Kim. A study of the formation and structure of small clusters of noble gases and sulfur hexafluoride formed in Laval nozzle molecular beams. PhD thesis, Northwestern University, 1981.
5. M. Kremmer and O. Okurounmu. Condensation of ammonia vapor during rapid expansion. Contract Report NONR 3963 (07) of the Office of Naval Research. Gas Turbine Laboratory, Massachusetts Institute of Technology, Cambridge, 1965.

6. P. P. Wegener and J.-Y. Parlange. Condensation by homogeneous nucleation in the vapor phase. *Naturwissenschaftern*, 57(11), 525–533, 1970.

7. K. Yano. Computer simulation on generation of nitrogen cluster (in Japanese). *RIKEN Rep.*, 51, 139–145, 1975.

8. I. Yamada, J. Matsuo, and N. Toyoda. Cluster ion beam processing. In *2002 International Conference on Ion Implantation Technology Proceedings*, eds. B. Brown, T. L. Alford, M. Nastasi, and M. C. Vella, 661–664. Institute of Electrical and Electronics Engineers, Piscataway, NJ, 2002.

9. Z. Insepov and I. Yamada. Direct Monte Carlo Simulation of supersonic conical nozzles. In *Extended Abstracts of the 3rd Workshop on Cluster Ion Beam Process Technology*, Kyoto, October 10–12, 2002, pp. 163–168.

10. J. Hautala, M. Gwinn, W. Skinner, and Y. Shao. Productivity enhancements for shallow junctions and DRAM applications using infusion doping. In *Proceedings of the 16th International Conference on Ion Implantation Technology*, eds. K. J. Kirkby, R. Gwilliam, A. Smith and D. Chivers, vol. 866, 174–177. AIP Conference Proceedings. American Institute of Physics, New York, 2006.

11. J. P. Dykstra. Ionizer for gas cluster ion beam formation. US Patent 6629508 B2, filed December 8, 2000.

12. M. E. Mack. Gas cluster ion beams for wafer processing. *Nucl. Instrum. Methods* B, 237, 235–239, 2005.

13. M. E. Mack. Ionizer and method for gas-cluster ion-beam formation. U.S. Patent US2006/0097185 A1, filed November 2006.

14. O. F. Hagena. Cluster beams form nozzle sources. In *Molecular Beams and Low Density Gas Dynamics*, ed. P. P. Wegener, 93–181. Marcel Dekker, New York, 1974.

15. T. D. Märk. Cluster ions: Production, detection and stability. *Int. J. Mass Spectrom. Ion Process.*, 79, 1–59, 1987.

16. K. Stephan, H. Helm, and T. D. Märk. Mass spectrometric determination of partial electron impact ionization cross sections of He, Ne, Ar, and Kr from threshold up to 180 eV. *J. Chem. Phys.*, 73(8), 3763–3778, 1980.

17. P. Scheier and T. D. Märk. Doubly charged argon clusters and their critical size. *J. Chem. Phys.*, 86(5), 3056–3057, 1987.

18. P. Scheier and T. D. Märk. Triply charged argon clusters: Production and stability (appearance energy and appearance size). *Chem. Phys. Lett.*, 136, 423–426, 1987.

19. O. Echt, D. Kreisle, and E. Recknagel. Dissociation channels of multiply charged van der Waals clusters. *Phys. Rev. A*, 38(7), 3236–3248, 1988.

20. F. Bottiglioni, J. Coutant, and M. Fois. Ionization cross section for H_2, N_2 and CO_2 clusters by electron impact. *Phys. Rev. A*, 6(5), 1830–1843, 1972.

21. T. D. Märk and O. Echt. Internal reactions and metastable dissociations after ionization of van der Waals clusters. In *Clusters of Atoms and Molecules II*, ed. H. Haberland, 154–182. Springer-Verlag, Berlin, 1994.

22. J. Gspann and K. Körting. Cluster beam of hydrogen and nitrogen analyzed by time of flight mass spectrometry. *J. Chem. Phys.*, 59, 4726–4734, 1973.

23. M. E. Mack, R. Becker, M. Gwinn, D. R. Swenson, R. P. Torti, and R. Roby. Design issues in gas cluster ion beamlines. In *Proceedings of the 14th International Conference on Ion Implantation Technology*, eds. B. Brown, T. L. Alford, M.Nastasi, and M. C. Vella, 665–668. Institute of Electrical and Electronics Engineers, Piscataway, NJ, 2002.

24. T. Seki, J. Matsuo, G. H. Takaoka, and I. Yamada. Generation of the large current cluster ion beam. *Nucl. Instrum. Methods B*, 206, 902–906, 2003.

25. K. Sattler, J. Mühlback. O. Echt, P. Pfau, and E. Recknagel. Evidence for Coulomb explosion of doubly charged micro clusters. *Phys. Rev. Lett.*, 47, 160–163, 1981.

26. D. Kreisle, O. Echt, M. Knapp, E. Recknagel, K. Leiter, T. D. Märk, J. J. Sáenz, and J. M. Soler. Dissociation channels for multiply charged clusters. *Phys. Rev. Lett.*, 56, 1551–1554, 1986.

27. O. Echt and T. D. Märk. Multiply charged clusters. In *Clusters of Atoms and Molecules II*, ed. H. Haberland, 183–220. Springer-Verlag, Berlin, 1994.

28. N. Toyoda, M. Saito, N. Hagiwara, J. Matsuo, and I. Yamada. Cluster size measurement of large Ar cluster ions with time of flight. In *1998 International Conference on Ion implantation Technology Proceedings*, eds. J. Matsuo, G. Takaoka, and I. Yamada, 1234–1237. Institute of Electrical and Electronics Engineers, Piscataway, NJ, 1999.

29. A. T. Forrester. *Large Ion Beams*. John Wiley & Sons, Hoboken, NJ, 1988.

30. I. Yamada, J. Matsuo, E. C. Jones, D. Takeuchi, T. Aoki, K. Goto, and T. Sugii. Range and damage distribution in cluster ion implantation. *Mater. Res. Soc. Symp. Proc.*, 438, 363–374, 1997.

31. J. Jortner. Level structure and dynamics of clusters. *Phys. Chem.*, 88, 188–201, 1984.
32. D. R. Swenson. Measurement of averages of charge, energy and mass of large, multiply charged cluster ions colliding with atoms. *Nucl. Instrum. Methods B*, 222, 61–67, 2004.
33. D. R. Swenson. Analysis of charge, mass and energy of large gas clusters ions and applications for surface processing. *Nucl. Instrum. Methods B*, 241, 599–603, 2005.
34. N. Toyoda and I. Yamada. Gas cluster ion beam equipment and applications for surface processing. *IEEE Trans. Plasma Sci.*, 36, 1471–1488, 2008.

5

Cluster Ion–Solid Surface Interaction Kinetics

There are unique interactions between cluster ion beams and surfaces that occur only with cluster ion beams and not with conventional ion beams. In this section, fundamental cluster ion beam interactions that were recognized during initial experimental studies and from molecular dynamics simulations are shown. Interactions of gas cluster ion beams (GCIBs) created by supersonic expansion of gas materials and relatively small polyatomic cluster ion beams are described.

5.1 Gas Cluster Ion Beams

The bombardment effects of monomer ions and large cluster ions are very different even at the same energy per atom, not only with respect to implant range, but also in terms of damage formation effects. It has already been shown that low-energy effects, lateral sputtering effects, and highly activated chemical effects are important characteristics of GCIB bombardment. To illustrate the difference between atomic ion and cluster ion bombardments, Figure 5.1 gives an example of simulations by SRIM [1] for 8 keV monomer ions and molecular dynamics (MD) for an 8 keV Ar_{2000} cluster ion. The comparison

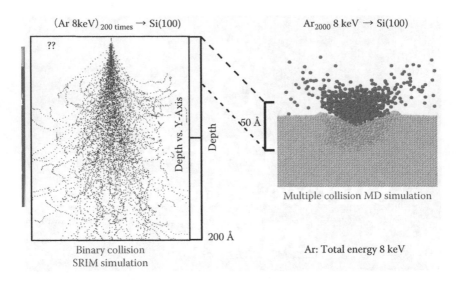

Figure 5.1 Comparison between monomer ion and cluster ion bombardment effects.

shows that even though the total acceleration energies are the same for both monomer and cluster ions at 8 keV, the range distributions are enormously different. Even though the total energies of both beams are the same, the actual bombardment energies are 8000 eV for the monomer ions and only 4 eV/atom for the 2000-atom cluster ion. Furthermore, Figure 5.1 indicates that not only the range distribution, but also the defect formation, is very different.

When an accelerated cluster ion hits a solid surface, it creates a small, highly compressed and heated "spike" region. The atoms of the target undergo multiple collisions with one another, resulting in a highly nonlinear collision cascade. Figure 5.2 shows, from molecular dynamics simulation, a momentum profile for a collision of an Ar cluster on a Si(100) target [2]. The snapshots represent the impact of a 688-Ar-atom cluster having 80 eV/atom after 0.2, 0.4, and 1.2 ps. When the cluster collides with the target, the kinetic energy of the cluster is transferred to the target isotropically and a symmetrical crater is formed. Within the region close to the crater, the atomic arrangement has become highly disordered. In the momentum profile, the length of the lines represents the mean kinetic energy, and the direction of the lines shows

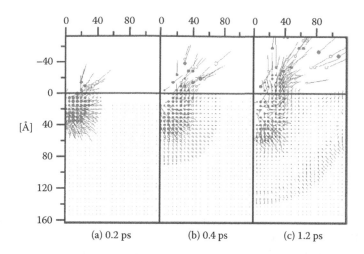

Figure 5.2 Momentum profile simulated by molecular dynamics for a collision of an Ar_{688} impact with an energy of 55 keV on a Si(100) target. The snapshots show (a) 0.2, (b) 0.4, and (c) 1.2 ps after the impact. Filled circles show the kinetic energy of the cluster atoms, and the open circles are for target atoms.

the momentum. These results indicate that many particles are sputtered with directions lateral to the trajectory of the impinging cluster. They also show that many atoms with large lateral momentum become present in the rim of the crater. The kinetic energy of these atoms is higher than 2 eV. Figure 5.2 also shows a shock wave that propagates outward from the impact region.

These phenomena lie well beyond the limits of the Sigmund linear collision theory [3]. Cluster impact involves electronic excitation of both the projectile and the target [4]. The impacted area experiences both high temperature and pressure transients: local temperatures can rise to between 10^4 and 10^5K, and pressures can rise to beyond a megabar (Mbar) (shown in Section 2.3 and Figure 2.19).

Figure 5.3 shows MD snapshots of Ar clusters of various sizes impacting on a Si surface. Each cluster has the same total incident energy of 20 keV; in other words, each has a different energy per atom. The results that are shown indicate that the amount and structure of damage are strongly dependent on cluster size. When cluster size is in the range of several hundreds to several thousands of atoms, large numbers

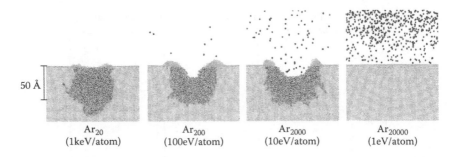

50 Å

| Ar$_{20}$ | Ar$_{200}$ | Ar$_{2000}$ | Ar$_{20000}$ |
| (1keV/atom) | (100eV/atom) | (10eV/atom) | (1eV/atom) |

Figure 5.3 MD snapshots of impacts of 20 keV Ar clusters of various sizes at 16 ps after impact.

of Si atoms are displaced spherically, and craterlike damage remains on the surface after the reevaporation of incident Ar atoms. When cluster size is as small as 20 atoms, craterlike damage is not formed, but a densely damaged region remains on the surface, with a deeper depth profile than that caused by Ar$_{200}$ or Ar$_{2000}$. On the other hand, when the cluster size is as large as 20,000 atoms and incident energy is as low as 1 eV/atom, the cluster does not penetrate the surface, but instead breaks up entirely on the surface. During this collisional process, some surface atoms are displaced, but the displacement distance is small, so that these displacements recover, and no damage remains on the surface after several picoseconds following the impact.

Figure 5.4 shows energy dependence of surface disruption caused by Ar$_{2000}$ impacts on a Si(100) surface at periods of 8 ps

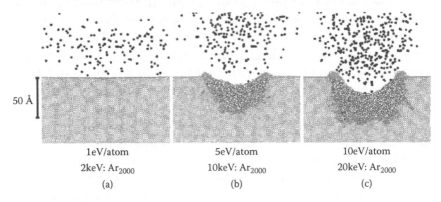

50 Å

1eV/atom	5eV/atom	10eV/atom
2keV: Ar$_{2000}$	10keV: Ar$_{2000}$	20keV: Ar$_{2000}$
(a)	(b)	(c)

Figure 5.4 MD snapshots of Ar$_{2000}$ clusters with various incident energies impacting on a Si(100) surface at 8 ps following impact.

after impact [5]. The total energy of Ar_{2000} was varied from 2 to 20 keV. When the total energy was 20 keV (10 eV/atom), the Ar_{2000} cluster penetrated into the Si surface and formed a hemispherical damaged area. Volume of the damaged region decreased with decreasing energy per atom. In the case of 2 keV energy (1 eV/atom), the surface was deformed at the first moment of the impact, but there was not enough energy to damage the Si crystal, and consequently no damage was left on the surface.

Figure 5.5 shows the mean depth dependence of dislocated Si atoms as functions of the incident energy per atom and the total acceleration energy, respectively [5]. The Ar cluster size was changed from 1 to 4000 atoms. In Figure 5.5a the energy per atom dependence of the damage depth due to Ar_{10} bombardment is similar to that by Ar_1 (monomer). For clusters of size larger than 10 atoms, the damage depth increases as the cluster size increases; however, it does not follow $E^{1/3}$ at energies below 10 eV/atom. In this situation, the cluster collapses on the solid surface, and no penetration occurs. From MD simulations, the energy that produces no damage on the surface is found to be in the range between 1 and 5 eV/atom.

Figure 5.5b shows total energy dependences of the mean depth of displacement. In the case of small clusters less than 200 atoms, there is no dependence on the cluster size, and the damaged depth increases with $E^{1/3}$. When a cluster has enough energy to penetrate into a substrate, its energy is transferred isotropically by multiple collisions, and as a result, a hemispherical damage area or craterlike trace is formed. Since the volume of the damaged area is almost proportional to the total acceleration energy, the damaged depth is proportional to $E^{1/3}$.

As for larger cluster impact, even if the total energy is as high as several keV, the incident energy per atom decreases as the cluster size increases. When cluster size is large enough, the cluster does not penetrate the surface of the target. In this case, the cluster stays on the surface, and most of the incident energy of the cluster is used to dissociate itself. The threshold energy per atom, where a cluster cannot penetrate the target and damage depth starts to obey the $E^{1/3}$ tendency, is about 10 eV/atom. This value is independent of the cluster size and

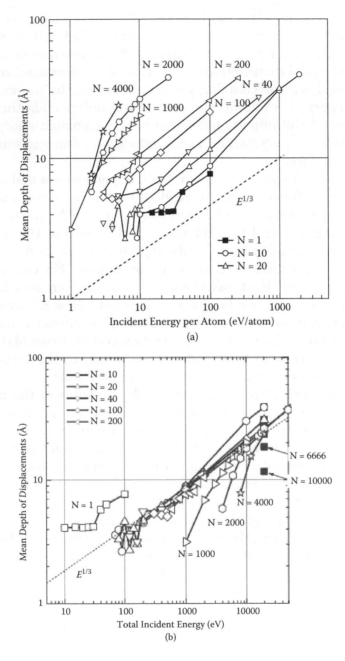

Figure 5.5 Mean depth of displacements produced by impacts of clusters of various sizes and energies (a) as a function of energy per atom and (b) as a function of total energy.

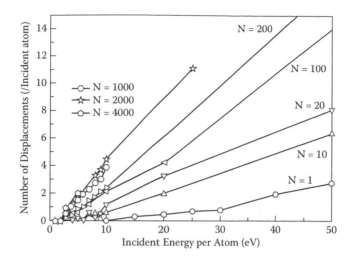

Figure 5.6 Incident energy per atom dependences of the number of displaced atoms for different cluster sizes.

is similar to the energy required to cause a knock-on atom on the target surface.

Figure 5.6 shows the number of displaced Si atoms produced per constituent atom of an Ar cluster ion versus the energy per atom. The results show that (1) there is a threshold incident energy to cause displacements in the target that depends on the cluster size, and (2) if the incident energy is larger than the threshold energy, the number of displaced atoms increases proportionally with the incident energy.

In order to experimentally evaluate damage layer thickness after higher cluster ion dose irradiations, crystalline Si substrates were irradiated with size-controlled Ar-GCIB, and the damage thickness was determined experimentally from ellipsometry measurements [6]. A two-layer model was used for the analysis, which assumed that oxide and amorphous layers were formed by the GCIB bombardment. The intensity ratio Ψ and phase difference Δ of the p and s waves could be identified. An increase in amorphous layer thickness is indicated by an increase of Ψ, and an increase in the oxide layer thickness is indicated by a decrease of Δ. From the Ψ and Δ behaviors, estimates could be made of the damage formation due to Ar-GCIB bombardment.

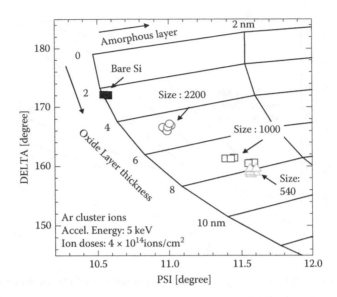

Figure 5.7 Cluster size dependence of Ψ and Δ after 5 keV Ar-GCIB irradiation with ion dose of 4×10^{14} ions/cm².

Si surfaces were bombarded by size-selected 5 keV Ar cluster ion beams to dose levels of 4×10^{14} ions/cm². Mean cluster sizes utilized were 540, 1000, and 2200 atoms/cluster, resulting in average energies of 9.2, 5.0, and 2.3 eV/atom, respectively. At the fixed 5 keV beam energy, both the oxide thickness and the amorphous layer thickness were found to decrease monotonically with increasing cluster size, that is, with decreasing energy of the cluster atoms. Figure 5.7 shows Ψ and Δ plots after Ar-GCIB irradiation of the Si.

Figure 5.8 shows cluster size dependence of the damage layer thickness in Si after Ar-GCIB irradiation with acceleration energy of 5 keV, plotted together with the number of displacement Si atoms by Ar cluster ion impact calculated by MD simulations. Ar cluster size was varied from 500 to 20,000 atoms. The cluster ion dose was 1×10^{14} ions/cm², and the irradiation angle was normal to the Si substrates. Both molecular dynamics simulation and experimental results showed good agreement, and the damage dropped around cluster size 1000. There was no damage or displacement in Si beyond Ar cluster size 3000.

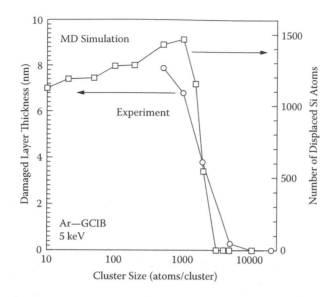

Figure 5.8 Cluster size dependence of the damage layer thickness in Si after Ar-GCIB irradiation with acceleration energy of 5 keV plotted together with the number of displacement Si atoms by Ar cluster ion impact as calculated by MD.

5.2 Polyatomic Ion Beams

As with large cluster ions, the bombarding characteristics of small cluster ions are also very different from those of monomer ions. Figure 5.9 shows molecular dynamics simulations of monomer (B_1), small cluster (B_{13} and B_{43}), and large cluster (B_{169}) ions impacting at 7 keV onto crystalline Si at 385 fs [7]. Figure 5.9a shows the typical monomer ion implantation effect. A collision cascade is created that causes dislodgement of atoms deep under the surface. The damage appears to be elongated, like a string extending below the penetration point. The small cluster with 13 atoms shown in Figure 5.9b collapses on impact with the surface, and the damage created exhibits features different from those associated with monomer implantation: the damage zone resembles a triangle, with a broader base and a shorter depth than the "string" appearance of the monomer. In the 169-atom cluster (Figure 5.9d), part of

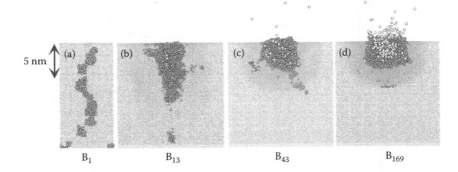

Figure 5.9 Molecular dynamic simulations of monomer (B_1), small cluster (B_{13} and B_{43}), and large cluster (B_{169}) ions impacting on crystalline Si at 7 keV after an elapsed time of 385 fs.

the cluster remains together as a unit on impact, while only a few isolated atoms break off and generate collision cascades. The penetration is shallower for these cascades, because of the lower energy per atom. In this case, the cluster itself remains essentially intact during impact, and a hemispherical "shell of damage" forms around the cluster as it stops. There is still a possibility of channeling, as the acceptance angle for axial and planar channeling increases as the atom energy decreases for small clusters, but the channeling observed in the simulations decreases with increasing cluster size, as expected.

For implantation by small clusters, when implantation energy is high enough, the incident B atoms penetrate into the target individually. This can be explained to be the result of negligible interactions between the cluster atoms. On the other hand, for impacts by low-energy clusters, the interactions between the cluster atoms, referred to as the multiple-collision effect, become significant. The results of these simulations indicate that clusterlike bombardment phenomena, the non-linear effects typical of cluster impact, occur already with clusters of 13 atoms. (This is discussed further in Figure 5.11.)

A fundamental question in cluster ion implantation is, how large must a cluster ion be in order to produce nonlinear cluster effects during the bombardment of a solid surface? Figure 5.10 shows the implant damage distributions resulting from model clusters of monomer, 2 atoms, 5 atoms, and 10 atoms [8, 9]. The results of these simulations indicate that

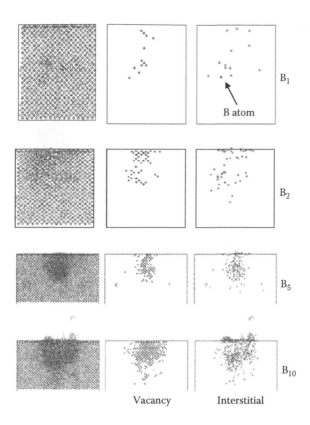

Figure 5.10 MD simulation of typical implant damage distributions for B_1, B_2, B_5, and B_{10} ions (acceleration energy: 500 eV/atom).

clusterlike bombardment phenomena, the nonlinear effects typical of cluster impact, begin to be observed with clusters containing five or more atoms. MD simulations of bombardment by even larger clusters have shown that the displacement damage increases with increasing cluster size, and that very large clusters, as in the case of GCIB, cause complete self-amorphization.

Figure 5.11 shows MD simulation depth profiles of implanted B atoms by boron monomer and cluster implantations at 200 eV/atom, 500 eV/atom, and 1 keV/atom. The results show that when the implant energy is as high as 1 keV/atom, the B_{10} and B_{18} clusters give almost the same implant profile as monomer (B_1) ions. As implant energy decreases, however, the difference of depth profiles between monomer and clusters becomes

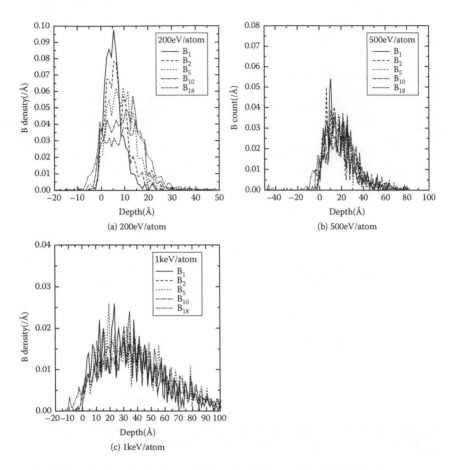

Figure 5.11 Depth profiles of implanted boron atoms.

significant. In the example of 200 eV/atom implant energy, B_{10} and B_{18} are implanted deeper than in the case of B_1, whereas B_2 gives a profile similar to that of B_1. These results indicate that the nonlinear collision processes by cluster ions depend not only on cluster size, but also on bombarding energy.

From the MD simulations, Figure 5.12a shows predicted mean implant depths (R_p) as functions of incident energy per atom for B monomers and for clusters with 2, 5, 10 and 18 B atoms. Figure 5.12b shows the R_p depth enhancement of the clusters relative to R_p of the monomers. The enhancement of R_p by multiple collisions that resulted only at the lowest incident energy can be explained to be a "clearing-way effect" in which

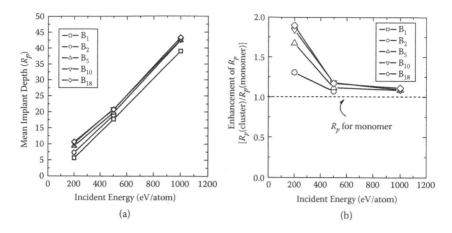

Figure 5.12 Incident energy dependence of (a) mean implant depth R_p and (b) enhancement ratio of R_p.

the first incident cluster atom displaces the target atoms at the near-surface region, and then the following cluster atoms do not lose their incident energies at the surface, and consequently penetrate more deeply.

The low-energy effect has been experimentally confirmed by using a decaborane ($B_{10}H_{14}$) implantation into a Si surface. A decaborane ion is actually a polyatomic molecular ion, not a gas cluster ion, but it does contain 10 boron atoms and 14 hydrogen atoms. When an ionized and accelerated decaborane molecule impacts upon a surface, the nature of this impact, which involves simultaneous impingement by 10 boron atoms, is essentially the same as what would occur in the impact of a cluster ion that might have been formed from 10 boron-containing molecules. Figure 5.13 shows boron distributions that resulted from implantations by 0.5 keV boron monomer ions and 5 keV decaborane ions [10]. The ion dose was 1×10^{14} ions/cm^2 for the B ions. For decaborane, the ion dose was one order of magnitude lower (1×10^{13} ions/cm^2) than that used for the B implants, because each decaborane ion implants 10 B atoms. The boron atom distribution from 5 keV decaborane ions is seen to be almost identical to that resulting with 500 eV boron monomer ions.

The MD simulations and experiments described above showed fundamentals of bombarding kinetics of cluster ions

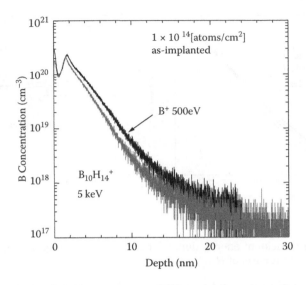

Figure 5.13 As-implanted B profiles of B and $B_{10}H_{14}$ implanted into Si. $B_{10}H_{14}$ dose implanted at 5 keV was 10^{13} ions/cm². B dose implanted at 500 eV was 10^{14} ions/cm².

on solid surfaces. The most significant result was that the bombarding kinetic energy of a cluster ion was confirmed experimentally to be roughly equal to the energy of the cluster divided by the number of atoms contained in the cluster. Another important result was the introduction of a shallow doping technology based on the self-low-energy effect of cluster ion bombardment.

Basic phenomena related to cluster size effects in nonlinear bombarding processes can also be observed by using secondary ion emission resulting from bombardment by cluster ions of different sizes. Experiments were conducted using mass analyzed C_n fragments produced from a C_{60} source [11]. Figure 5.14 shows the velocity dependence of secondary ion yield with C_n cluster ions (n = 1, 7, 9, 11, 60) from a Au surface. The secondary ion yield was normalized to the cluster size (i.e., the number of secondary ions produced by one atom with the same velocity). From the figure, the secondary ion yield of small clusters (C_7) was almost the same as that of the carbon monomer ions, which followed the nuclear stopping power curve of a carbon ion in Au calculated with transport of ions

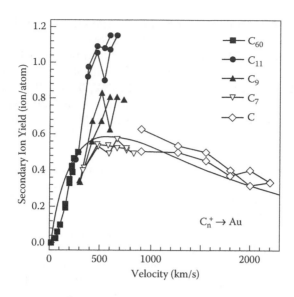

Figure 5.14 Secondary ion yields by carbon cluster ions C_n (n = 1, 7, 9, 11, 60) from a Au surface.

in matter (TRIM). At this cluster size, there was no difference in the ion emission behavior of monomer ions and small cluster ions. However, an enhanced secondary ion yield was observed starting from a cluster size of nine. At the ion velocity of 600 km/s, the secondary ion yield with C_{11} was twice as high as in the cases of C and C_7. This result agreed well with the results shown in Figure 5.11; that is, the cluster effect that is nonlinear starts at a cluster size between seven and nine atoms.

References

1. J. F. Ziegler, J. P. Biersack, and M. D. Ziegler. *SRIM: The Stopping and Range of Ions in Matter.* SRIM Co., Chester, MD, 2008.
2. T. Aoki. Molecular dynamics simulation of cluster ion impact on solid surface. PhD thesis, Kyoto University, 2000.
3. P. Sigmund. Sputtering by ion bombardment: Theoretical concepts. In *Sputtering by Particle Bombardment I*, ed. R. Behrisch. Springer-Verlag, Berlin, 1981.

4. R. J. Beuhler and L. Friedman. Large cluster ion impact phenomena. *Chem. Rev.*, 86, 521–527, 1986.

5. T. Aoki and J. Matsuo. Molecular dynamics simulation of surface modification and damage formation by gas cluster ion impact. *Nucl. Instrum. Methods B*, 242, 517–519, 2006.

6. N. Toyoda, S. Houzumi, T. Aoki, and I. Yamada. Experimental study of cluster size effect with size-selected cluster ion beam system. *Mater. Res. Soc. Symp. Proc.*, 792 R10.8.1–R10.8.6, 2004.

7. I. Yamada, J. Matsuo, N. Toyoda, and T. Aoki. Gas cluster ion beam processing. Professional Division Report MC-97-7 (in Japanese). Japan Society of Electric Engineering, Tokyo, November 18, 1997.

8. I. Yamada. 20 years history of fundamental research on gas cluster ion beams and current status of the applications to industry. In *Proceedings of the 16th International Conference on Ion Implantation Technology*, vol. 866, 147–154. AIP Conference Proceedings. American Institute of Physics, New York, 2006.

9. T. Aoki and J. Matsuo. Molecular dynamics simulations of boron cluster implantation. In *7th Workshop on Cluster Ion Beam Process Technology and Quantum Beam Process Technology*, Tokyo, November 6–7, 2006, pp. 65–70.

10. K. Goto. A study of sub-0.1μm CMOS process technology (in Japanese). PhD thesis, Tohoku University, 1998.

11. N. Toyoda. Nano-processing with gas cluster ion beams. PhD thesis, Kyoto University, February 1999.

6

Cluster Ion Beam Sputtering

6.1 Lateral Sputtering

When a cluster impacts upon a solid surface, the impinging energy of the cluster is deposited into a local area, resulting in the formation of a crater characterized by a central depression surrounded by a protruding rim. Since this characteristic crater formation does not occur in the case of monomer ion impact, a different sputtering mechanism and a different angular distribution of ejected atoms are associated with the cluster ion bombardment.

Figure 6.1 shows the angular distributions of Cu atoms sputtered at normal incidence by 20 keV Ar monomer ions and by 10 and 20 keV Ar cluster ions. The average cluster size and the ion dose were 2000 atoms/cluster and 1×10^{17} ions/cm^2, respectively [1]. The Cu film used as the sputter target in this study was initially polycrystalline with grains approximately 400 nm in diameter and 15 nm in height. This Cu surface became effectively fully smoothed by the Ar cluster ions after an ion dose of 3×10^{15} ions/cm^2, which was just a few percent of the total ion dose employed in the experiment. Therefore, the initial surface roughness in this case was considered to have had little influence on the measured angular distributions. In the case of 20 keV Ar monomer ions, the measured angular

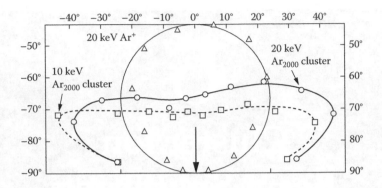

Figure 6.1 Angular distributions of Cu atoms sputtered at normal incidence by 20 keV Ar monomer ions and by 10 and 20 keV Ar cluster ions.

distribution formed the circle shown in the center of Figure 6.1. This circular distribution closely followed the cosine law. The angular distribution due to the Ar cluster ions, on the other hand, showed a shape completely different from the cosine distribution. With the reduction of incident energy to 10 keV, the angular distribution became even more flattened.

Angular distributions of sputtering produced by monomer ion bombardment have been topics of many studies because they give information that is important for the understanding and analysis of sputtering phenomena. The collision cascade theory of sputtering, which is generally considered to give a good quantitative description of the monomer ion sputtering process, predicts a cosine distribution of the sputtered atoms. Publications by Sigmund have presented good explanations of the theory and mechanisms of physical sputtering due to particle impact [2, 3]. Previous investigators had examined angular distributions of sputtered atoms using theoretical and experimental approaches. In early experiments, Wehner showed that the particles sputtered from metal single crystals were ejected preferentially in directions aligned with closest packing in the crystal lattices, i.e., in the (100) orientation in face-centered cubic (fcc) and (111) orientation in the case of body-centered cubic (bcc) [4]. Wehner and Rosenberg reported sputtering from polycrystalline targets by Hg⁺ ions at energies from 100 to 1000 eV [5]. The angular distributions remained "under cosine" at energies below 1000 eV, but with a tendency to approach closer to cosine distributions at increasing ion energy. Mo and Fe showed more pronounced tendencies for ejection toward lateral directions than did Ni or Pt.

Tsuge and Esho reported angular distributions of sputtered atoms from polycrystalline metal targets measured for 0.5 and 1.0 keV Ar$^+$ ions and compared the distributions with those resulting from a Si single-crystal target [6]. Figure 6.2 shows the change in angular distribution with ion energy at normal incidence for Au, NiFe, and Si at 0.5 keV (curve 1) and at 1.0 keV (curve 2). The metals were polycrystalline with more or less preferred (111) orientations, while the Si target was a (100) single crystal. As the ion energy increased from 0.5 to 1.0 keV, the amount of atom ejection along the target normal direction increased. Consequently, for (a) sputtered Au and (b) evaporated NiFe, preferential ejection in the close-packed directions was weakened with increasing ion energy. For (c) Si bulk, the distribution changed from greatly under cosine at 0.5 keV to more nearly cosine at 1.0 keV. This tendency appears to be consistent with the experimental results shown by Wehner and Rosenberg.

Andersen and coworkers conducted precise experiments using targets of ultra-high-vacuum (UHV)-deposited polycrystalline Cu, Pt, and amorphous Ge. Measurements of angular distributions were performed for irradiations by Ar at high energies ranging from 1.25 to 320 keV onto substrates that were surrounded by a cold shield kept at the liquid nitrogen (LN$_2$) temperature. The results showed that all targets yielded angular distributions more outward peaked than the cosine distribution predicted by the collision cascade theory. The germanium results were well fitted by a $\cos^n \Theta$ distribution, with n varying from 1.3 at the lowest energy to 1.6 at higher energies [7]. Wucher and Reuter investigated the energy dependences of angular distributions of metals and alloys under Ar ion bombardment at normal incidence. They showed that at higher bombarding energy (2 keV) the distributions exhibited a $\cos^3 \Theta$ behavior, but at low bombarding energy (250 eV) pronounced emission occurred at an oblique angle [8].

Yamamura studied sputtering at energies close to threshold values [9, 10]. He recognized that atoms sputtered in

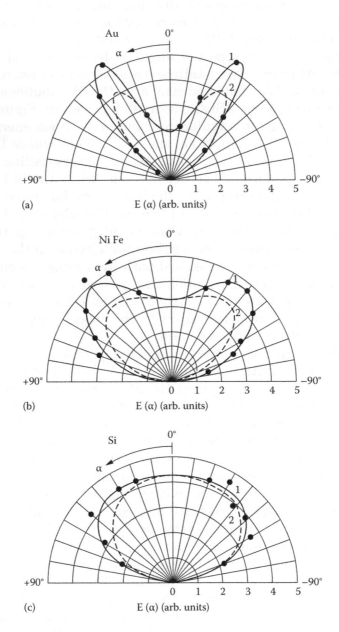

Figure 6.2 Change in angular distribution with ion energy at normal incidence. Curve 1, 0.5 keV; curve 2, 1.0 keV.

the near-threshold energy regime were generated by a few collisions, and all collision events took place only at the topmost or second atomic layers. Yamamura classified those collision events and analyzed them for different combinations of projectile ion energies and substrate materials. He also reviewed 137 papers published during the period from 1957 to 1988 and presented a compilation of the angular distributions of sputtered atoms at normal incidence and at oblique incidence for various combinations of incident ions and target atoms. He explained that at very low energies, close to threshold values, the low-energy incident projectile does not produce a well-developed collision cascade, and as a result, the polar diagram distributions are heart shaped. As projectile energies are increased beyond threshold, a cosine distribution is obtained, and at still higher energies, the distributions become over cosine, i.e., more outwardly peaked. For the overcosine distributions, there are two possible explanations. One possibility is that the surface introduces an asymmetry causing the recoil flux in the cascade to be anisotropic at the surface. Another possible explanation involves a so-called missing-plane model, in which the distributions are strongly influenced by outward scattering due to neighboring atoms at the topmost layer [11].

Figure 6.1 shows that there is a pronounced difference in the angular distributions produced by Ar monomer ion and Ar cluster ion bombardments. This difference has been explained by molecular dynamics simulations of the highly nonlinear collision processes occurring during cluster ion impact in which (as was shown in Figure 6.2) kinetic energy is transferred to surface atoms in an isotropic direction, thereby resulting in craterlike damage. During the formation of a crater, some atoms on the edge of the crater leave the surface in lateral directions. The characteristic sputtering distribution of cluster ion irradiation is associated with this craterlike damage formation action [12, 13].

Figure 6.3 shows measured incidence angle dependence of sputtering yields of Cu and Ag due to bombardment by 20 keV Ar cluster ions relative to the known distribution behavior caused by Ar monomer ions. The ion dose and the average

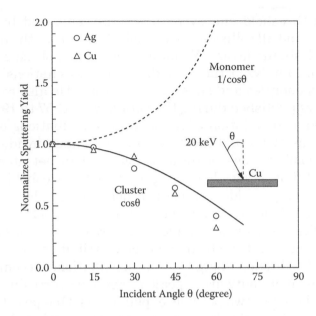

Figure 6.3 Incident angle dependence of the sputtering yields of Cu and Ag resulting from 20 keV Ar cluster ions.

cluster size were 5×10^{15} ions/cm^2 and 2000 atoms/cluster, respectively. The measured sputtering yields were normalized to those at normal incidence. The well-known 1/cos θ dependence of sputtering yields due to monomer ions is plotted for reference. A radical difference between the cluster and monomer characteristic behaviors is obvious. In the case of both the Cu and Ag targets, the normalized sputtering yields of 20 keV Ar cluster ions decreased in proportion to cos Θ. When the incidence angle was 0°, the penetration was deepest, and most of the kinetic energy of the clusters was transferred to the target atoms as a result of nonlinear collision processes. In the case of oblique incidence, the energy carried by the cluster ions was only partially transferred to the target because many atoms from the cluster were ejected or reflected from the surface with large kinetic energies. Since the kinetic energy deposited perpendicularly onto the target was proportional to cos Θ, the sputtering yield followed this curve.

In the case of monomer ions, sputtering phenomena, including the penetration depth, density of deposited energy, damaged area, and sputtering yield, depend on the angle of

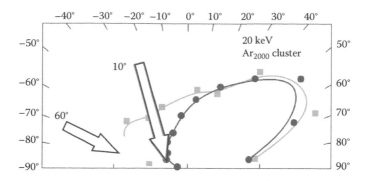

Figure 6.4 Angular distribution of sputtered atoms at incident angles of 10° and 60°.

incidence of the bombarding ions. In cluster ion sputtering, the incidence angle also has a strong, but different, influence. Figure 6.4 shows the angular distribution of Cu atoms sputtered by 20 keV Ar cluster ions incident at oblique angles ranging from 10° to 60°. These distributions are seen to be different from those shown in Figure 6.1, resulting at 0° incidence. Even at an oblique incidence angle of only 10°, the majority of the sputtered particles were distributed in the forward direction. No significant change in the angular distribution was found as the angle of incidence was changed from 10° to 60°. The result shows that the localization of energy density occurs during an oblique impact, and that this energy is preferentially released from one side.

6.2 Physical Sputtering

Sputtering by cluster ion beams involves high sputtering yields and strong surface-smoothing effects. Figure 6.5 shows an atomic force microscope (AFM) image of a Au target irradiated by an Ar cluster ion beam through an ultrafine mesh mask. The acceleration energy, cluster size, and ion dose were 20 keV, 2000 atoms/cluster, and 1×10^{16} ions/cm^2, respectively. Several square areas, which are seen to have been reduced in height, were the regions sputtered by the Ar cluster ions. The bottoms of the irradiated areas became highly smoothed

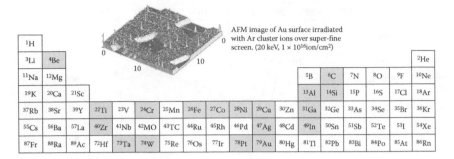

AFM image of Au surface irradiated with Ar cluster ions over super-fine screen. (20 keV, 1×10^{16} ion/cm^2)

¹H																	²He
³Li	⁴Be											⁵B	⁶C	⁷N	⁸O	⁹F	¹⁰Ne
¹¹Na	¹²Mg											¹³Al	¹⁴Si	¹⁵P	¹⁶S	¹⁷Cl	¹⁸Ar
¹⁹K	²⁰Ca	²¹Sc															
³⁷Rb	³⁸Sr	³⁹Y	²²Ti	²³V	²⁴Cr	²⁵Mn	²⁶Fe	²⁷Co	²⁸Ni	²⁹Cu	³⁰Zn	³¹Ga	³²Ge	³³As	³⁴Se	³⁵Br	³⁶Kr
⁵⁵Cs	⁵⁶Ba	⁵⁷La	⁴⁰Zr	⁴¹Nb	⁴²MO	⁴³TC	⁴⁴Ru	⁴⁵Rh	⁴⁶Pd	⁴⁷Ag	⁴⁸Cd	⁴⁹In	⁵⁰Sn	⁵¹Sb	⁵²Te	⁵³I	⁵⁴Xe
⁸⁷Fr	⁸⁸Ra	⁸⁹Ac	⁷²Hf	⁷³Ta	⁷⁴W	⁷⁵Re	⁷⁶Os	⁷⁷Ir	⁷⁸Pt	⁷⁹Au	⁸⁰Hg	⁸¹Tl	⁸²Pb	⁸³Bi	⁸⁴Po	⁸⁵At	⁸⁶Rn

Figure 6.5 AFM image of a Au target irradiated by an Ar cluster ion beam through an ultrafine mesh mask and periodic table identifying materials that have been tested for cluster ion beam sputtering.

relative to the surface of the unprocessed area, which was protected by the mask.

Sputtering yields of Ar cluster ions in various materials (shown in the periodic table of Figure 6.5) have been determined from the slopes of measured linear experimental relationships between ion dose and resulting sputter depth. Unless this linearity exists, the sputtering yields cannot be defined. Figure 6.6 shows the ion dose dependence of sputter depth of various elements due to 20 keV Ar cluster ion beam bombardment [14]. For these determinations, average cluster size was 2000 atoms/cluster. The target materials were films of Ag, Cu, Au, W, Zr, and Ti deposited onto Si. The cluster ion dose was varied from 8×10^{14} to 8×10^{15} ions/cm^2. Fairly good linearity of the dose dependence can be seen for the different materials.

Figure 6.7 shows the sputtering yields of various materials due to 20 keV Ar cluster ions. (The figure also includes data from SF$_6$ clusters, which produce chemical etching in addition to sputtering, as will be discussed in Section 6.3.) In order to validate the experimental procedure for determining sputtering yields due to cluster bombardment, similarly measured sputtering yields for 20 keV Ar monomer ions were compared with known reference values [14]. The monomer sputtering yields showed periodic changes related to the atomic numbers of the target materials. This tendency occurs because sputtering yield is inversely proportional to the binding energy of the atoms. Consequently, sputtering yields of the atoms in the Ib group (Cu, Ag, and Au) are higher than those in the IVa group (Ti, Zr) or VIa group (W). The reason is that the binding energy becomes low whenever the d shell is filled by electrons, which in turn results in an increase in the sputtering yield [15]. In these experiments, the measured sputtering yields of monomer ions showed good agreement with the reported values, thus validating the experimental method.

Compared with the sputtering yields measured with Ar monomer ions at the same acceleration energy, sputtering yields obtained with Ar cluster ions were approximately one order of magnitude higher. For example, the sputtering yield of Cu with Ar cluster ions (65.5 atoms/ion) was 12 times higher than that with Ar monomer ions (5.3 atoms/ion). Moreover, the sputtering yield with Ar cluster ions followed the same periodic changes as those observed with monomer ions. Hence, the sputtering yields of other materials with Ar cluster ions can be roughly estimated from the reference values for Ar monomer ions.

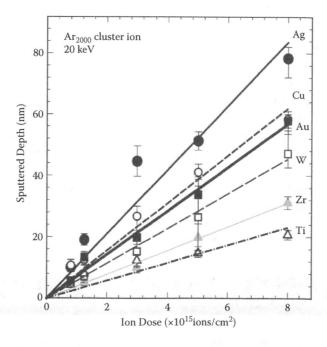

Figure 6.6 Ion dose dependence of sputtered depth with 20 keV Ar cluster ions for various materials.

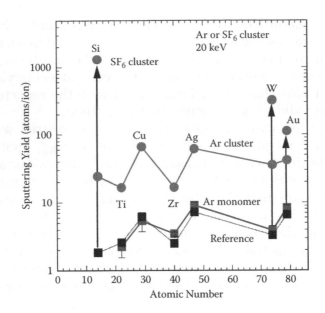

Figure 6.7 Sputtering yields of various materials due to 20 keV Ar monomer ions and Ar cluster ions.

Figure 6.8 shows the dependence of the sputtering yields of Cu and Ag on the acceleration energy of Ar cluster ions. The ion dose for these experiments was 5×10^{15} ions/cm², and the average cluster size was 2000 atoms/cluster. The acceleration energy ranged from 10 to 25 keV. The sputtering yields increased almost linearly with acceleration energy over this range. There was a threshold energy for sputtering at 6 keV, where on average each cluster atom had a kinetic energy of approximately 3 eV. In this case, collisions near the surface were dominant, and all the energy was efficiently deposited onto the surface. Thus, the target surface atoms can be sputtered at an energy per cluster atom lower than the threshold value for sputtering by monomer ions.

6.3 Reactive Sputtering

The sputtering yields with Ar cluster ions for various materials are one or two orders of magnitude higher than those of

Ar monomer ions at the same energies. In the case of reactive gas species, however, additional enhancement of sputtering yields due to the chemical reactions can be expected. Figure 6.9 shows sputtering yields of Si, SiC, W, and Au due to Ar monomer ions, Ar cluster ions, and SF_6 cluster ions, all at energies of 20 keV. The average size of the Ar and SF_6 cluster ions was 2000 atoms. In comparing sputtering of W and Au, which have similar atomic masses, the sputtering yield of Au due to Ar cluster ions (42 atoms/ion) was only slightly higher than that of W (35 atoms/ion). The same tendency was found with Ar monomer ions. On the other hand, the sputtering yield of W due to SF_6 cluster ions was 320 atoms/ion, which was almost three times higher than that of Au (112 atoms/ion) due to the same SF_6 cluster ions. The reason is that the tungsten atoms reacted with fluorine dissociated from SF_6 clusters; as a result, the sputtering yield was enhanced by the production of volatile WF_x materials. Enhancement of the sputtering yields with SF_6 cluster ions was also observed in the case of compound materials such as SiC.

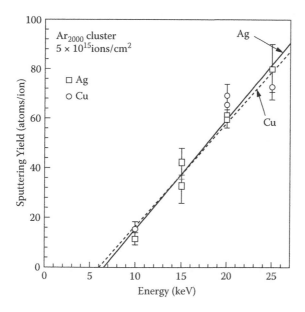

Figure 6.8 Energy dependence of the sputtering yield with Ar cluster ions.

Similarly, the sputtering yield of Si by Ar cluster ions (24 atoms/ion) was enhanced 55 times by chemical reaction with SF_6 cluster ions (1300 atoms/ion). Since the average size of the SF_6 clusters was about 2000 atoms, the SF_6 cluster ions showed high reaction probability with Si, even though a SF_6 molecule is stable at room temperature. An explanation is that, in the etching process, the following sequence of events is believed to occur: absorption of the etchant on the surface, formation of a volatile product, and evaporation or ejection of the product from the surface. It has been reported that chemical reactions between SF_6 molecules and Si are stimulated by energetic Ar monomer ion bombardments as a result of SF_6 molecules being dissociated by the energetic ion bombardment and the released fluorine atoms then being able to react with Si atoms [16]. The fragmentation of a cluster and the dissociation of a SF_6 molecule occur simultaneously when a SF_6 cluster ion collides with a Si or W surface. Consequently, SiF_x or WF_x, which are volatile compounds, are produced. SiF_x and WF_x are physically sputtered by energetic ions or are thermally evaporated into the vacuum. Thus, the sputtering yield increases as a consequence of the production of volatile compounds promoted by SF_6 cluster ion bombardment. This process was

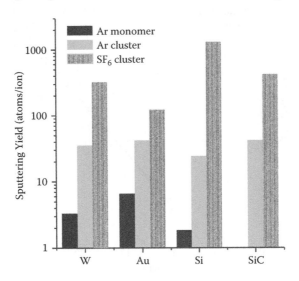

Figure 6.9 Sputtering yields of Si, SiC, W, and Au by Ar monomer ions, Ar cluster ions, and SF_6 cluster ions at 20 keV.

confirmed by measuring the mass spectra during the irradiation of the Si target with a 25 keV SF_6 cluster beam using a residual gas analyzer [17].

The angular distributions of sputtered atoms are expected to be strongly affected by these chemical reactions. When the projectile ion reacts with the target atoms, compound products are generated on the surface as a consequence of the chemical reactions, and these increase or decrease the sputtering yields. The angular distributions of atoms sputtered by SF_6 cluster ion beams have been measured on both reactive and nonreactive targets, specifically, tungsten and gold films, respectively. The SF_6 reacted with tungsten and gave rise to both physical and reactive sputtering effects, whereas the Au target suffered only physical sputtering effects from the SF_6 cluster ions. The atomic masses of tungsten (184) and gold (197) are almost the same. The mass ratio of an impinging ion relative to the target atoms is significant in physical sputtering phenomena. Therefore, in the case of the W and Au targets, the effects of chemical reactivity on the angular distributions can be compared without considering the mass ratio of primary ion to target atom.

Figure 6.10 shows the angular distribution of the Au and W atoms sputtered by SF_6 cluster ions. The acceleration energy of the cluster ions was 20 keV, and the incidence angles were $0°$ and $10°$. The angular distribution of sputtered Au due to cluster ion sputtering at normal incidence showed a flattened profile, which was the same as that of Cu due to Ar cluster ions (as shown in Figure 6.1). In addition, sputtered Au atoms were distributed in the forward direction at an incidence angle of $10°$. Since no chemical reaction occurred on the Au surface with SF_6 cluster ions, these angular distributions were induced only by physical sputtering. From these results, it appears that the lateral sputtering at normal incidence and the marked change of angular distribution at near normal incidence are inherent to physical sputtering due to any cluster ion.

In contrast to the situation with Au, the angular distribution of W sputtered with SF_6 cluster ions almost followed the cosine law at normal incidence. As discussed before, tungsten reacted with fluorine and produced reaction products such as WF_6, a volatile material that was then thermally evaporated from the surface isotropically. As a result, the angular distribution followed a cosine law. Even at an oblique impact, the angular distribution was almost isotropic. This was completely different from the behavior of Ar cluster ions on a Cu surface or that of SF_6 cluster ions on a Au surface at oblique impact, which showed asymmetric distribution to the forward side of the incident direction.

However, the number of sputtered particles at high angles (near to the surface) was larger than would be expected from the cosine law (shown as a circle in Figure 6.10). The difference of the yield between the as-measured and the cosine distribution were also plotted (shown as a dotted line). This additional yield showed an undercosine distribution similar to that of Cu with Ar cluster ions. This could be attributed to the physical sputtering effects of SF_6 cluster ions on W. Chemical reaction was a dominant etching process at this energy level. However, the contribution from physical sputtering with reactive cluster ions showed the same distribution as with the Ar cluster ions. Therefore, both physical and chemical sputtering effects were present in the SF_6 cluster impact, and the angular distribution became isotropic when chemical reactions were dominant.

Various reactive cluster ion beams, such as CF_4, C_2H_6, C_3H_8, CHF_3, CH_2F_2, Cl_2, and O_2, have been generated by mixing with He gas. Sputtering yields by these beams are found to increase with acceleration voltage and are usually much higher than the yields by Ar monomer ions [18].

6.4 Surface Smoothing

Surface smoothing is an inherent characteristic of cluster ion beam bombardment, which usually does not occur with monomer ions. In order to study the difference in surface-smoothing effects caused by monomer and cluster ions, Cu films were deposited onto Si substrates by sputtering and were irradiated at ambient temperature and normal incidence with Ar monomer and cluster ion beams. The average cluster size was 2000 atoms/cluster. The ion doses were 1.2×10^{16} ions/cm^2 with Ar monomer ions and 8.0×10^{15} ions/cm^2 with Ar cluster ions.

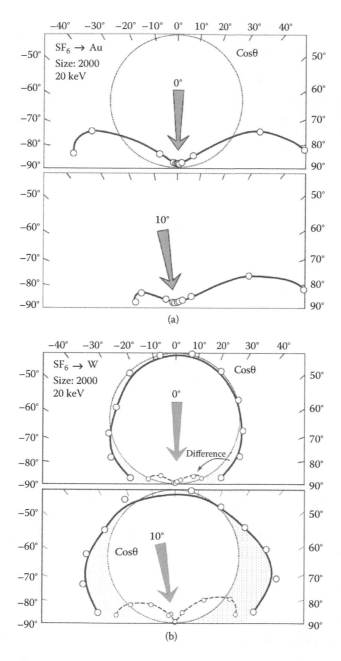

Figure 6.10 Angular distribution of (a) Au atoms sputtered by SF_6 cluster ions and (b) W atoms sputtered by SF_6 cluster ions.

The surfaces were examined by AFM. The scan area of the AFM was either 1 × 1 µm or 10 × 10 µm. Figure 6.11 shows the AFM images of the Cu surfaces. The initial Cu surface (Figure 6.11a) had many grains typically about 400 nm wide and 15 nm in height, and the average roughness was 5.8 nm. Following Ar monomer ion bombardment (Figure 6.11b), there were still many grains on the surface, and small hillocks caused by the energetic ion bombardment were observed. The average roughness of this surface was 4.9 nm, which means that the surface roughness was only slightly improved by the 20 keV Ar monomer ions at normal incidence.

When the Cu target was irradiated by an Ar cluster ion beam (Figure 6.11c), the average roughness was reduced to 1.3 nm, which was only a little more than one-fifth of the initial roughness. Considering that the cluster ion dose was only two-thirds that of the monomer ions, the Cu surface was obviously smoothed more effectively by the Ar cluster ions than by the Ar monomer ions. In monomer ion sputtering, it has been reported that the anisotropy of the sputtering yield depends on the presence of facets or grain boundaries. In Figure 6.11c, no grains are observed on the surface, and there seems to be little dependence of sputtering yield with cluster ions on grain boundaries or facets. In the same way as in the case of Cu targets, other metals, semiconductors, and insulator surfaces have been smoothed effectively by using cluster ions.

Figure 6.12 shows the ion dose dependence of the average roughness of the Cu surfaces bombarded at normal incidence with 20 keV Ar cluster ion beams with doses from 8×10^{14} to 8×10^{15} ions/cm^2. AFM images at each ion dose are shown. As can be seen in the figure, the average roughness decreased monotonically with increasing ion dose from the initial value of 6 nm to 1.3 nm at an ion dose of 8×10^{15} ions/cm^2. In the case of monomer ion irradiation at normal incidence, the surface roughness increased with ion dose as a result of erosion or bubble formation inside the target. With Ar cluster ion irradiation, the roughness of the Cu surface decreased to a saturation value at a dose of about 3×10^{15} ions/cm^2, and then no subsequent roughening occurred with further increase of dose.

Ripple formation at glancing incidence by cluster ions is quite different from that by monomer ions. Figure 6.13 shows

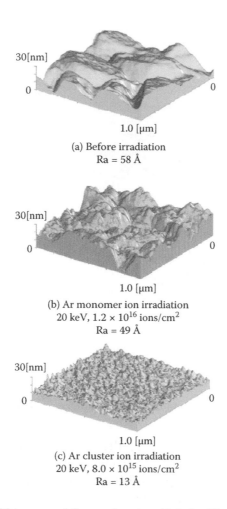

(a) Before irradiation
Ra = 58 Å

(b) Ar monomer ion irradiation
20 keV, 1.2×10^{16} ions/cm^2
Ra = 49 Å

(c) Ar cluster ion irradiation
20 keV, 8.0×10^{15} ions/cm^2
Ra = 13 Å

Figure 6.11 AFM images of Cu surface irradiated with Ar monomer and cluster ions.

AFM images of Cu surfaces irradiated with Ar cluster ions at several incident angles (0°, 15°, 30°, 45°, and 60°) [19]. The average size of the Ar cluster ions was 2000 atoms/cluster. The acceleration energy and the ion dose were 20 keV and 5×10^{15} ions/cm^2, respectively. The initial Cu surface shown in Figure 6.13a exhibited an average roughness of 6 nm in a 1×1 μm area. In the case of the incidence angle of 0° (Figure 6.13b), the Cu surface was smoothed, and its average roughness became 1 nm. On increasing the incidence angles to 15° and 30° (Figure 6.13c and d), distortions began to appear

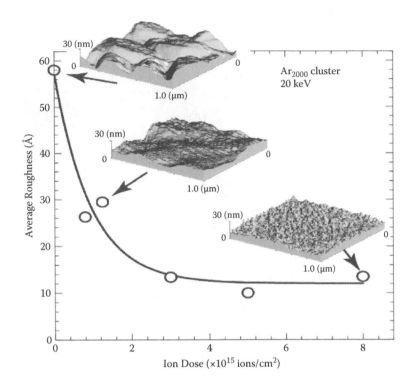

Figure 6.12 Ion dose dependence of the surface roughness of Cu with Ar cluster ions.

on the surface. Nevertheless, although the surface-smoothing effect was less strong with increasing incidence angle, the surface roughness was improved from the initial value of 6 nm to 1.9 and 4.4 nm at incidence angles of 15° and 30°, respectively.

When the incidence angle exceeded 45°, however, the surface roughness increased, and ripples whose wave vector was parallel to the incident direction appeared. The wavelength of the ripple produced by a 60° impact was about 0.2 μm. The average roughness values at incidence angles of 45° and 60° were 7.1 and 11.3 nm, respectively, which are higher than the initial value of 6 nm. Similar ripple formations at glancing angles were also observed on silver surfaces.

The increase in the surface roughness at oblique impact was not due to the reduction of the sputtered depth. The relation between the sputtered depth and the average roughness was measured and is shown in Figure 6.14. If the

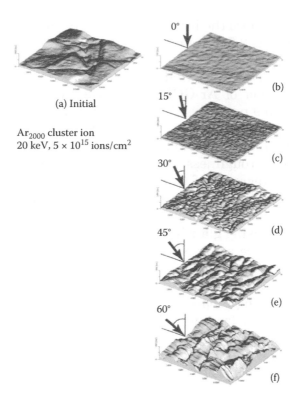

(a) Initial

Ar$_{2000}$ cluster ion
20 keV, 5 × 10^{15} ions/cm^2

Figure 6.13 Incident angle dependence of the surface roughness with Ar cluster ions.

surface-smoothing effects at an oblique incidence and at normal incidence were the same, the data points should lie on the same line as at normal incidence, but this was not the case. With an incidence angle of 60°, Cu was sputtered to a depth of about 10 nm, and its average roughness was 13 nm, which was five times higher than that for normal incidence at the same sputtered depth. Clearly, then, a surface-roughening mechanism exists at oblique incidence.

Reduced roughness of Au surfaces by increasing the incidence angle of Ar cluster ion bombardment to beyond 60° has been reported [20]. Figure 6.15 shows average surface roughness and depths of material removed as measured on Au films exposed to 20 keV 1 × 10^{16} ions/cm^2 Ar-GCIB irradiations at incidence angles of 0°, 30°, 60°, and 85°. The sputtered depths were measured from etching steps by using a surface contact profilometer. Before the irradiations, the surface roughness of the Au films was 2.5 nm, shown as a dotted line in Figure 6.15. The deepest sputtering occurred at normal incidence, and the depth of sputtering decreased with increasing incidence angle. After irradiation of Ar-GCIB at 0°, the surface roughness decreased to 0.9 nm. When the incidence angle was increased to 30°, the average surface roughness was 1.5 nm (still better than initially), but when the incidence angle was increased to 60°, surface roughness increased dramatically to 9.4 nm. With further increase of the incidence angle to 85°, the surface roughness again dropped markedly to 1.3 nm, but at this angle very little material was removed. Thus, it can be concluded that by using glancing-angle GCIB irradiation, surface smoothing can be achieved without removing large amounts of material.

Figure 6.16 shows the influence of angle of incidence upon surface smoothing produced by physical and reactive sputtering. Au and W surfaces were irradiated with 20 keV SF_6 cluster ions at incidence angles of 0° and 60°. The ion dose and the average cluster size were 7×10^{15} ions/cm^2 and 2000 atoms/cluster, respectively. The nonreactive Au target suffered only

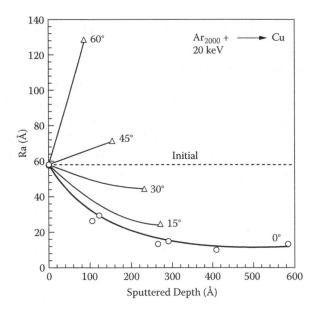

Figure 6.14 Incident angle dependence of surface roughness.

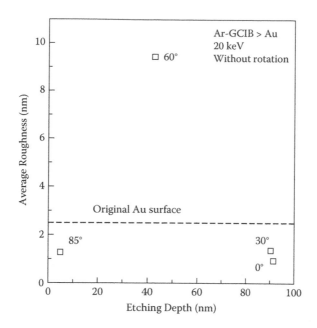

Figure 6.15 Average roughness of Au sputtered by 20 keV Ar-GCIB at incident angles of 0°, 30°, 60°, and 85°.

physical sputtering due to the SF_6 cluster ions, whereas SF_6 reacted with the tungsten and produced both physical and reactive actions. The initial average roughnesses of the Au and W surfaces were 7.3 and 19 nm, respectively. The figure shows AFM images of the Au and W surfaces initially (a and d) and after irradiations at incidence angles of 0° (b and e) and 60° (c and f).

In the case of Au, the surface roughness was improved from 7.3 to 2.4 nm by normal incidence bombardment. At an incidence angle of 60°, the surface roughness became 8.5 nm, which was higher than the initial value; there was a roughening effect at oblique impact with SF_6 cluster ions that was the same as that of Cu with Ar cluster ions at oblique impact. Surprisingly, ripples whose wave vectors were parallel to the incident direction were formed at an incident angle of 60° (these were also observed at high angles of Ar cluster impact, as was shown in Figure 6.13).

In contrast, the surface roughness of W was not improved as much as in the case of Au due to SF_6 cluster ions at normal

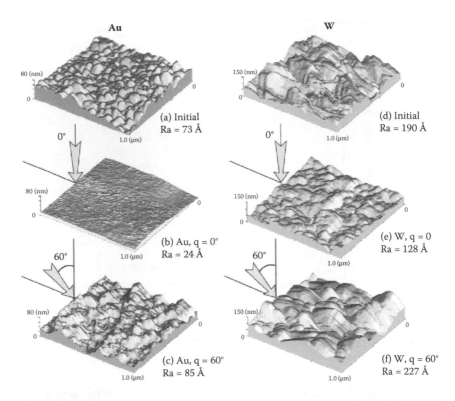

Figure 6.16 Surface morphology of Au and W surface irradiated with SF$_6$ cluster ions.

incidence. The average roughness was slightly reduced from 19 to 12.8 nm after the SF$_6$ cluster ion irradiation. When the sample was tilted to an incidence angle of 60°, the surface roughness increased from 19 to 22.7 nm, and no ripple formation was observed. The reactive sputtering caused by SF$_6$ cluster ions on W did not produce effective smoothing effects; however, the enormously high sputtering yields due to reactive sputtering can be useful for high-speed etching and smoothing by combining with physical sputtering by Ar.

Effects of surface smoothing of MgF$_2$, Si, and SiO$_2$ substrates at incidence angles up to 85° by reactive GCIB processing have been reported [21]. As was seen previously in the case of Cu sputtering by Ar cluster ions, surface roughnesses increased as bombardment angles were increased from normal incidence to about 60°; then at angles greater than 60°, the

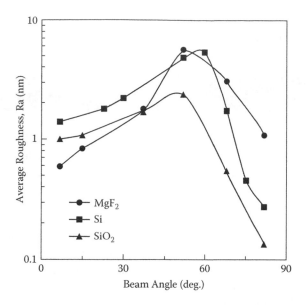

Figure 6.17 Average roughness dependence on beam incident angle by SF_6 cluster beam irradiation at 30 keV and 1×10^{15} ions/cm^2.

roughness values decreased substantially on all the materials. As shown in Figure 6.17, the average surface roughness of MgF_2 after bombardment at 82° was 1 nm, which was higher than the roughness after bombardment at normal incidence. For Si and SiO_2, the average surface roughnesses resulting at 82° incidence were 0.27 and 0.14 nm, respectively, substantially smaller than the roughnesses resulting on these materials at normal incidence.

In order to understand the surface-smoothing behavior of cluster ion beams at glancing-angle incidence, molecular dynamics simulations were carried out using 20 keV Ar cluster ions incident on Si substrates [22]. The results indicated that when the perpendicular component of the incident energy of a cluster impacting on a plane surface at a large angle of incidence is less than several electron volts per atom, the incident cluster slides on the surface without creating the crater-like damage typically observed at normal incidence. However, when a cluster impacts on an irregular surface feature at a low glancing angle (80° from surface normal), the cluster causes multiple collisions at the irregular surface structure,

which leads to dynamic deformation by sputtering and produces surface smoothing.

References

1. N. Toyoda, H. Kitani, N. Hagiwara, T. Aoki, J. Matsuo, and
   I. Yamada. Angular distributions of the particles sputtered
   with Ar cluster ions. *Mater. Chem. Phys.*, 54, 262–265, 1998.
2. P. Sigmund. Theory of sputtering. I. Sputtering yield of amorphous and polycrystalline targets. *Phys. Rev.*, 184(2), 383–416,
   1969.
3. P. Sigmund. Mechanisms and theory of physical sputtering by
   particle impact. *Nucl. Instrum. Methods B*, 27, 1–20, 1987.
4. G. K. Wehner. Controlled sputtering of metals by low-energy
   Hg ions. *Phys. Rev.*, 102(3), 690–704, 1956.
5. G. K. Wehner and D. Rosenberg. Angular distribution of sputtered material. *J. Appl. Phys.*, 31, 177–179, 1960.
6. H. Tsuge and S. Esho. Angular distribution of sputtered atoms
   from polycrystalline metal targets. J. Appl. Phys., 52(7), 4391–
   4395, 1981.
7. H. H. Andersen, B. Stenum, T. Sørensen, and H. J. Whitlo.
   Angular distribution of particles sputtered from Cu, Pt and Ge
   targets by keV Ar+ ion bombardment. *Nucl. Instrum. Methods
   B*, 6, 459–465, 1985.
8. A. Wucher and W. Reuter. Angular distribution of particles
   sputtered from metals and alloys. *J. Vac. Sci. Technol.*, A6(4),
   2316–2318, 1988.
9. Y. Yamamura and J. Bohdansky. Few collisions approach for
   threshold sputtering. *Vacuum*, 35(121), 561–571, 1985.
10. Y. Yamamura, Y. Mizuno, and H. Kimura. Angular distribu-
    tions of sputtered atoms for low-energy heavy ions, medium
    ions and light ions. *Nucl. Instrum. Methods B*, 13, 393–395,
    1986.
11. Y. Yamamura, T. Takiguchi, and H. Tawara. Data compilation
    of angular distributions of sputtered atoms, 1–281. Research
    Report NIFS-DATA-1. National Institute for Fusion Science of
    Japan, Nagoya, Japan, 1990.
12. T. Aoki, J. Matsuo, Z. Insepov, and I. Yamada. Molecular dynam-
    ics simulation of damage formation by cluster ion impact. *Nucl.
    Instrum. Methods B*, 121, 49–52, 1997.

13. Z. Insepov, I. Yamada, and M. Sosnowski. Sputtering and smoothing of metal surface with energetic gas cluster beams. *Mater. Chem. Phys.*, 54, 234–237, 1998.
14. J. Matsuo, N. Toyoda, M. Akizuki, and I. Yamada. Sputtering of elemental metals by Ar cluster ions. *Nucl. Instrum. Methods B*, 121, 459–463, 1997.
15. N. Laegreid and G. K. Wehner. Sputtering yields of metals for Ar^+ and Ne^+ ions with energies from 50 to 600 eV. *J. Appl. Phys.*, 32, 365–369, 1961.
16. D. J. Oostra, A. Haring, A. E. de Vries, F. H. M. Sanders, and G. N. A. van Veen. Etching of silicon by SF_6 induced by ion bombardment. *Nucl. Instrum. Methods B*, 13, 556–560, 1986.
17. N. Toyoada, H. Kitani, N. Hagiwara, J. Matsuo, and I. Yamada. Surface smoothing effects with reactive cluster ion beams. *Mater. Chem. Phys.*, 54, 106–110, 1998.
18. Y. Shao, M. D. Tabat, C. K. Olsen, and R. MacCrimmon, Gas cluster ion beam etching process for Si-containing and Ge-containing materials. U.S. Patent US8513138 B2, filed August 20, 2013.
19. H. Kitani, N. Toyoda, J. Matsuo, and I. Yamada. Incident angle dependence of the sputtering effect of Ar-cluster-ion bombardment. *Nucl. Instrum. Methods B*, 121, 489–492, 1997.
20. N. Toyoda and I. Yamada. Surface roughness reduction using gas cluster ion beam for Si photonics. Presented at 5th International Symposium on Advanced Science and Technology of Silicon Materials (JSPS Si Symposium), Kona, HI, November 10–14, 2008.
21. E. Bourelle, A. Suzuki, A. Sato, T. Seki, and J. Matsuo. Sidewall polishing with a gas cluster ion beam for photonic device applications. *Nucl. Instrum. Methods B*, 241, 622–625, 2005.
22. T. Aoki and J. Matsuo. Molecular dynamics simulations of surface smoothing and sputtering process with glancing-angle gas cluster ion beams. *Nucl. Instrum. Methods B*, 257, 645–648, 2007.

7
Cluster Ion Implantation

Cluster ion implantation by large gas cluster ions and by poly-atomic cluster ions is a promising new area of opportunity for both research and practical applications such as semiconductor doping. The low-energy effect is the primary advantage. However, because of the low charge-to-mass ratio and the surface interactions of large aggregates of atoms at the same time, several other unique implantation features can also be achieved. Prospective advantages of cluster ion implantation include extremely shallow doping, mass-independent doping, $E^{1/3}$ depth dependence of implanted atoms, self-amorphization, reduced transient enhanced diffusion (TED), and absence of end-of-range (EOR) damage effects.

7.1 Gas Cluster Ion Implantation: Infusion Doping Process

Gas cluster ion implantation, called infusion doping, was introduced by Epion Corporation in the United States. The physics of infusion doping have been studied and are relatively well understood, and extensive development of both equipment and processes has been performed [1]. Figure 7.1 shows a conceptual sketch of the infusion processes. Gas cluster ion beam (GCIB) consists of Ar gas clusters as hosts into which

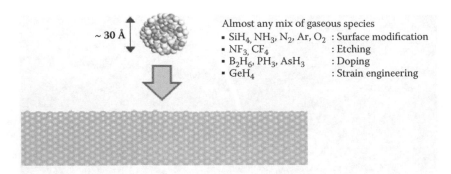

Figure 7.1 Infusion: gas cluster ion beam processing.

fractional amounts of dopant-containing gases such as B_2H_6, PH_3, or AsH_3 are added. The intense thermal spike produced at cluster impact causes these molecular gas species contained in the clusters to undergo energetic dissociation, thereby allowing their solid components (i.e., B, P, or As) to infuse into the target surface by an atomic mixing action that occurs within the thermal transient region.

In Figure 7.2 a secondary ion mass spectroscopy (SIMS) profile of B in silicon after 5 keV GCIB infusion doping using 1% B_2H_6 in Ar is compared with the profile produced by conventional 500 eV B^+ ion implantation. A junction depth (Xj) of 12 nm at the B concentration of 1×10^{18}/cm^3 was achieved with an abruptness of <2.5 nm/decade. The abruptness and total absence of channeling in the case of infusion are characteristic of this process, which depends on thermal transients rather than ballistic penetration. In the case of the conventional B implantation process, significant channeling can be seen, resulting in an Xj at 37 nm.

Mass-independent doping, in which the depth of dopant distribution by GCIB infusion is independent of the dopant species mass, is a unique characteristic of the infusion process [2]. Figure 7.3 shows SIMS profiles of simultaneous B and Ge infusion at 5 keV resulting from mixing B_2H_6 and GeH_4 gases into Ar gas clusters. The ratio of the retained doses is determined by the gas mix used for cluster formation, and thus the doping ratios, or stoichiometry, of the resulting compound materials such as $SiGe_xB_y$ can be controlled by mass flow controllers or the ratio of gas species premixed in the source bottle.

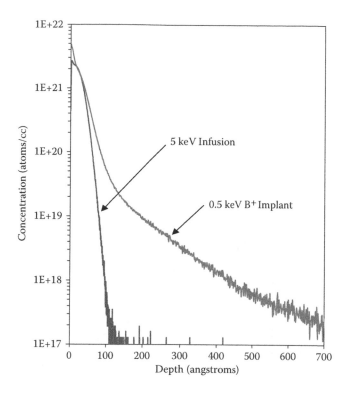

Figure 7.2 SIMS profiles from B infusion doping at 5 keV and B+ implantation at 500 eV.

Figure 7.4 shows the relationship of junction depth measured at $1 \times 10^{18}/cm^3$ versus infusion doping energy over the energy range of 2.5–60 keV [2]. For comparison, Figure 7.4 also includes junction depths resulting from traditional B+ implantation, BF_2^+ implantation [3], and $B_{18}H_{22}$ polyatomic cluster ion implantation [4], which will be discussed further in the next section. In conventional ion implantation processing, ion penetration is a ballistic action, and thus it results in an essentially linear relationship between the ion acceleration energy and the resulting depth of the implant. The slope of this relationship is mass dependent. On the other hand, for the infusion implantation, the hemispherical volume of the intermixing of the doping species and the substrate is determined by the collective energy of the cluster. Consequently, the depth follows a very distinct energy to the one-third (1/3) power dependence.

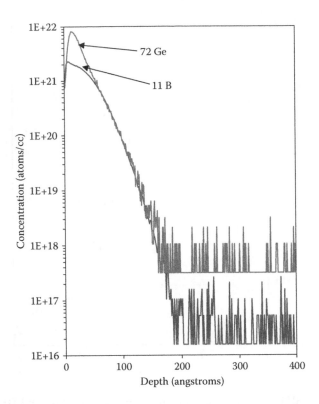

Figure 7.3 SIMS profiles of simultaneous infusion of B_2H_6 and GeH_4 into Si.

Upon impact with the substrate, the cluster locally heats a volume of Si generating a transient thermal spike (TTS), while the several thousand atoms in the cluster infuse into the surface with the megabar pressures that are generated over the few picoseconds of the impact. The TTS propagates in three dimensions and is quickly quenched. This situation is fundamentally different from conventional ion implantation in which the energy dissipates primarily along the ion trajectory (as shown in Figure 2.19 and elsewhere in molecular dynamics (MD) simulations). The effects of this TTS can be seen in Figure 7.5, where wafers that had been implanted by conventional 500 eV boron ion implantation were subsequently subjected to Ar GCIB exposures at various energies from 2.5 to 30 keV [1]. Note the flattening of the B profiles in Figure 7.5, thereby making them more boxlike in shape. Taking the Xj

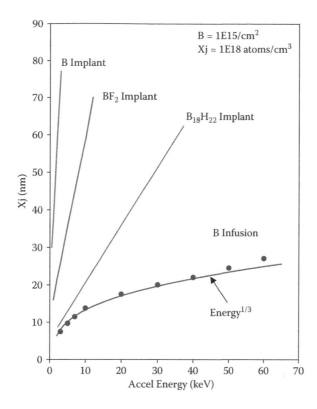

Figure 7.4 Doping depth as a function of beam acceleration energy for B, BF_2, $B_{18}H_{22}$, and GCIB. The junction depth measured at $1 \times 10^{18}/cm^3$.

values at $1 \times 10^{19}/cm^3$ after redistribution versus the Ar cluster energy, it can be seen that the data follow the same energy to the 1/3 power relationship shown in Figure 7.4.

Because GCIB infusion implantation employs the thermal transients created by cluster impact to introduce dopant atoms into a target surface, it is not subject to the same self-sputtering limits that are associated with conventional ion implantation. During conventional implantation by monomer or molecular ions, the surface being implanted is subject to continuous sputtering by those ions, thereby causing atoms to be lost from the surface. For shallow doping and high dopant concentrations, competition between the implantation and sputtering actions can establish a saturation condition in which the number of dopant atoms being lost from the surface due to sputtering matches the rate of arrival of new dopant ions. This

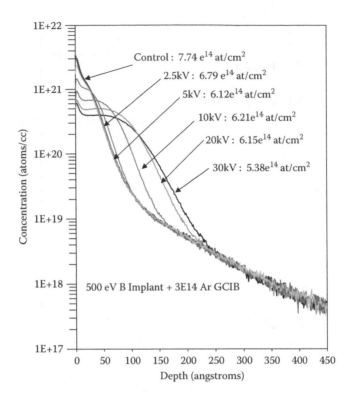

Figure 7.5 Transient thermal spike localized heating effects on ion-implanted boron dopant movement after various Ar gas cluster processing, making it more boxlike in shape.

same situation does not occur in the case of GCIB infusion until much higher dopant concentration levels are reached. Figure 7.6 shows an example of retained B dose in Si versus GCIB dose for an infusion doping process performed at 60 keV [5]. The retained B can be seen to have increased linearly with the process dose up to a level of beyond 10^{17} B atoms/cm².

7.2 Polyatomic Cluster Ion Implantation

Starting from the earliest development of production ion implantation technology, many different boron compounds, such as B_2H_6, $B_2N_3H_6$, B_2S_3, BF_3, BBr_3, BI_3, BCl_3, HBO_2, and $B_{10}H_{14}$,

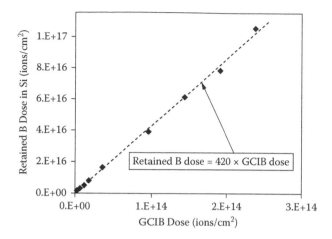

Figure 7.6 Retained B dose versus GCIB dose.

have been evaluated as candidate source materials for achieving higher B^+ ion currents [6]. BF_3 has generally been considered to be the most effective choice. However, in more recent times, as integrated circuit device structures progressed toward ever-smaller dimensions, very challenging requirements developed for not only higher B ion currents, but also for high currents at ion energies lower than the realizable lower limits of practical implantation equipment. Polyatomic implantation using entire large molecular species such as $B_{10}H_{14}$ was recognized to offer a solution for the problem of achieving ultrashallow junctions. Since the time when the first p-channel metal-oxide-semiconductor field-effect transistors (p-MOSFETs) with shallow junctions were successfully fabricated by using decaborane molecules [7], researchers have recognized the importance of polyatomic implantation for fundamental understanding and for practical applications. Implantation of different sizes of B polyatomic clusters, such as B_1, B_2, B_5, B_{10}, and B_{18}, shown in Figure 7.7, has been studied in detail [8–11].

An important issue is whether TED can be suppressed by decaborane ion implantation. TED is important for ultrashallow junction formation: when ions are implanted into the substrate, some substrate atoms are displaced from their normal sites, and some dopants are introduced into nonsubstitutional positions. For short-duration thermal annealing, the diffusion

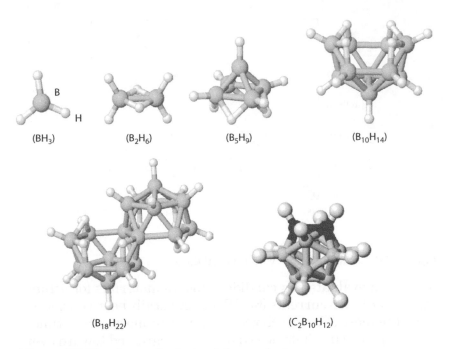

Figure 7.7 Structures of polyatomic boron clusters.

is enhanced over traditional diffusivity values, leading to significant spreading of the original dopant profile [12, 13].

The first experimental data on the energy dependence of dopant diffusion and its annealing characteristic are shown in Figure 7.8. The SIMS dopant profiles of boron, implanted at 5, 3, and 2 keV, before and after annealing at 900°C and 1000°C for 10 s by rapid thermal annealing, showed that shallow junction formation was possible, and that TED after the annealing at 900°C was essentially suppressed in these low-energy ranges. The thermal diffusion that occurred at 1000°C depended upon the dopant density and was a consequence of the very high concentration of boron in these very shallow implantations [7, 14, 15].

To determine the relationship between TED and $B_{10}H_{14}$ ion implantation, researchers investigated the effect of implant energy on the diffusion properties [16]. The diffusion distance was defined as the deviation between as-implanted and annealed profiles at B concentration of 3×10^{17} atoms/cm^3. Figure 7.9 shows this diffusion distance obtained from the

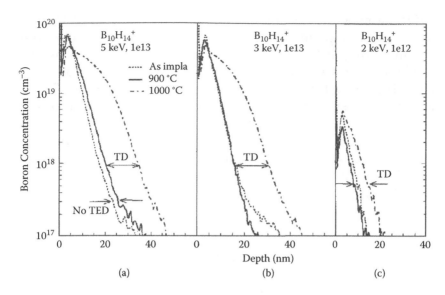

Figure 7.8 SIMS analyses of B concentration after $B_{10}H_{14}$ implantation at (a) 5 keV, (b) 3 keV, and (c) 2 keV before and after 10 s RTA at 900°C and 1000°C.

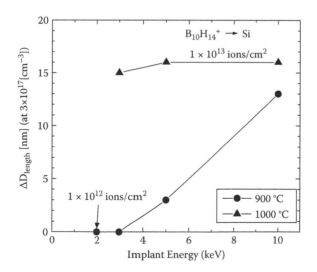

Figure 7.9 Dependence of boron diffusion distance on implantation energy.

SIMS profiles during annealing at 900°C and 1000°C as a function of implant energy. The figure demonstrates that for annealing at 900°C, TED was reduced by decreasing the implant energy. On the other hand, for annealing at 1000°C, the diffusion distance did not depend on implant energy. These results suggest that the mechanism of B diffusion is different at the two annealing temperatures.

Figure 7.10 shows the energy dependence of diffusivity enhancements (ratio of observed diffusivity to equilibrium diffusivity: D_B/D_B^*). Equilibrium boron diffusivities are given by the following equations [12]:

$$D_B^* = D_B^0 + D_B^+ \ (\text{cm}^2/\text{s})$$

$$D_B^0 = 0.278\exp(-3.946 \times 10^4/T)$$

$$D_B^+ = 0.23\exp(-3.946 \times 10^4/T)$$

For annealing at 900°C, the large enhancement value of 70 associated with 10 keV implantation resulted from the occurrence of TED. However, the enhancement value decreased with decreasing implant energy, and the TED enhancement

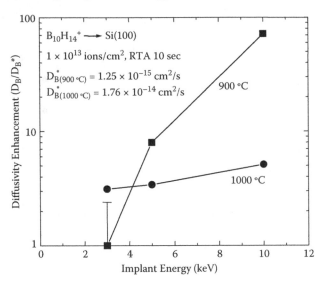

Figure 7.10 Dependence of diffusivity enhancement (D_B/D_B^*) on implantation energy.

was not observed when the incident energy was below 3 keV. This result indicates that low-energy $B_{10}H_{14}$ ion implantation successfully suppressed the TED during annealing at 900°C. For annealing at 1000°C, however, the enhancement values were almost constant with energy and were small, approximately 3 to 4. These results suggest that the mechanism of TED at 3 keV is not the same as that at 10 keV.

A possible explanation for the suppression of TED with decreasing implantation energy during 900°C annealing is that the damage is concentrated in a region very close to the surface. In such a case, the surface is believed to act as a sink for interstitials. Consequently, the interstitials introduced by ion implantation are recombined and annihilated at the surface, thereby suppressing the TED [17]. Figure 7.11 shows the observed experimental relationship between diffusivity enhancement and implanted B dose. At 3 keV, the enhancement is seen to be only weakly dose dependent for annealing at 900°C or 1000°C. At 900°C, diffusivity is very small, and the amount of boron diffusion could not be measured by SIMS for any of the implant doses. During 1000°C annealing, the amount of TED does depend weakly on implant dose, probably

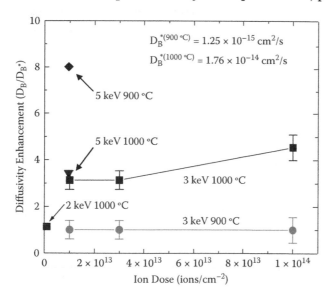

Figure 7.11 Dependence of diffusivity enhancement (D_B/D_B^*) on implantation dose.

because the number of defects near the surface has become saturated and the implanted region of substrate becomes amorphous for any dose greater than 1×10^{13} ions/cm^2. Figure 7.11 also shows diffusivity enhancement due to annealing at 1000°C for the sample implanted at 2 keV and 1×10^{12} ions/cm^2 and the diffusivity enhancement resulting from annealing at 900°C and 1000°C for the samples implanted at 5 keV with the dose of 1×10^{13} ions/cm^2.

Figure 7.12 shows the observed dependence of the number of disordered atoms on ion dose. $B_{10}H_{14}$ ions were implanted at 3 and 20 keV. The total numbers of resulting disordered atoms were obtained by integration over the surface peaks in the Rutherford backscattering and channeling spectra. The density of disordered atoms on the unirradiated surface, due to surface distortion and intrinsic vacancies, is shown as a reference baseline in Figure 7.12. In the case of $B_{10}H_{14}$ ion implantation at 20 keV, the number of disordered atoms saturated at about 6×10^{16} atoms/cm^3 for a dose of 5×10^{14} ions/cm^2. On the other hand, with 3 keV implant, the disordered atoms saturated at about 1.6×10^{16} atoms/cm^3 for a dose of 1×10^{14} ions/cm^2. Because the projected range (R_p) decreases with decreasing implant energy, the amorphous layer becomes shallower and the number of disordered atoms decreases, which

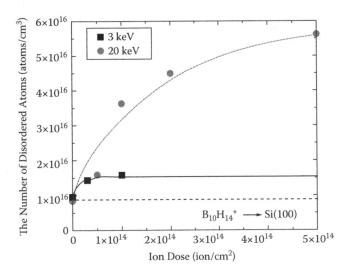

Figure 7.12 Dependence of the number of disordered atoms on ion dose.

means that the threshold dose required to amorphize the layer decreases with implant energy. In the case of $B_{10}H_{14}$ ion implantation at 3 keV, an amorphous layer was formed near the surface at a dose of 1×10^{13} ions/cm^2. Therefore, as was indicated in Figure 7.11, in the case of low-energy implantation, the enhancement during annealing at 900°C and 1000°C depended only weakly on implant dose. On the other hand, in the case of high-energy implantation, the implanted region, at a dose of 1×10^{13} ions/cm^2, was not fully amorphized, and large numbers of interstitial Si atoms remained in the substrate such that TED was consequently very extensive.

Differences in annealing behavior between cluster or monomer ion implantation have been studied by computer simulation of implantation of B_{10} with subsequent rapid thermal annealing (RTA). For this simulation, a combination of two methods, molecular dynamics and Monte Carlo, was used, because the annealing process takes seconds and is therefore inaccessible for MD simulation alone [18]. Figure 7.13a shows a view of a Si target at 8 ps after B_{10} impact at 2.5 keV. After obtaining the initial positions of implanted B atoms and Si substrate atoms by MD simulation, the as-implanted (MD) atomic positions were transferred to a Metropolis Monte Carlo (MMC) code to simulate an RTA process at 1050 K. Figure 7.13b shows the evolution of dopant trajectories during the RTA process. Clearly, almost all B trajectories remain in the vicinity of the amorphized substrate zone.

The simulation results show that a large area of the substrate is amorphized by the B cluster implantation. The implanted boron atoms are trapped within the amorphized area for the whole annealing time. The atomic movement of boron within the amorphized area is much easier than that for equilibrium diffusion, and therefore the resulting diffusion coefficients are high. This fast B diffusion in Si has no relation to TED because it occurs primarily in the x-y plane. Since both TED and thermal diffusion are almost perfectly suppressed by 3 keV implantation and by 900°C annealing, it is expected that ultrashallow junctions can be formed under these conditions.

Polyatomic cluster implantation using octadecaborane ($B_{18}H_{22}$) and carborane ($C_2B_{10}H_{12}$), which is more thermally stable, has also been reported [19]. Promising results for

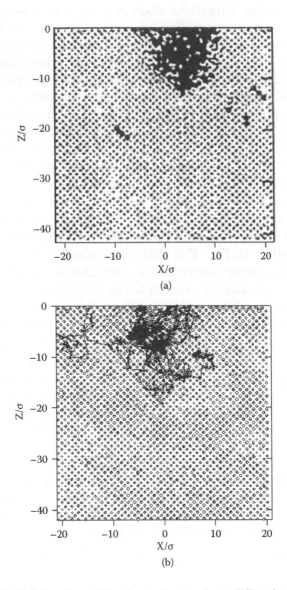

Figure 7.13 (a) Side view of B_{10} cluster ion implanted Si substrate by MD simulation. (b) B trajectories during RTA after cluster implantation by MMC simulation. $\delta = 2.0951$ Å, length unit in Stillinger–Weber potential for Si.

p-MOSFET fabrication—such as self-amorphization effects, undamaged smooth surface formation, smooth amorphous–crystal interface formation, diffusionless annealing, higher carrier activation effects, higher on-current characteristics, small-threshold voltage fluctuation, and superior suppression of short channel effects—were reported in comparison with traditional ion implantation processes [20]. Detailed characteristics of the integrated circuit (IC) devices are discussed in Chapter 9.

References

1. J. Hautala, J. Borland, M. Tabat, and W. Skinner. Infusion doping for USJ formation. In *4th International Workshop on Junction Technology (IWJT'04)*, 50–53. Institute of Electrical and Electronics Engineers, Piscataway, NJ, 2005.
2. J. Hautala, M. Gwinn, W. Skinner, and Y. Shao. Productivity enhancements for shallow junctions and DRAM applications, using infusion doping. In *Proceedings of the 16th International Conference on Ion Implantation Technology*, vol. 866, 174–177. AIP Conference Proceedings. American Institute of Physics, New York, 2006.
3. J. O. Borland, T. Matsuda, and K. Sakamoto. Shallow and abrupt junction formation: Paradigm shift at 65–70 nm. *Solid State Technol.*, 45(6), 83–94, 2002.
4. M. Tanjyo, T. Nagayama, N. Hamamoto, S. Umisedo, Y. Koga, N. Maehara, T. Matsumoto, N. Nagai, F. Ootsuka, A. Katakami, K. Shirai, T. Watanabe, H. Nakata, M. Kitajima, T. Aoyama, T. Eimori, Y. Nara, Y. Ohji, K. Saker, W. Krull, D. Jacobson, and T. Horsky. Cluster ion implantation for beyond 45nm node novel device applications. In *Extended Abstracts of the International Workshop on Junction Technology 2008 (IWJT '08)*, Shanghai, China, pp. 55–57.
5. W. Skinner, M. Gwinn, J. Hautala, and T. Kuroi. Infusion processing for advanced transistor manufacturing. In *2006 IEE/SEMI Advanced Semiconductor Manufacturing Conference*, Boston, May 22–24, 2001, pp. 214–218.
6. R. G. Wilson and D. M. Jamba. Comparison of sources of boron, phosphorus, and arsenic ions. *Appl. Phys. Lett.*, 22, 176–179, 1973.

7. K. Goto, J. Matsuo, T. Sugii, H. Minakata, I. Yamada, and T. Hisatsugu. Novel shallow junction technology using decaborane ($B_{10}H_{14}$). *IEDM Tech. Dig.*, 435–438, 1996.

8. J. E. Edward. Ion implantation system and method. PCT WO2011/056515 A2, filed October 25, 2010.

9. M. I. Current. Shallow and high-dose implantations: Atomic, molecular and cluster ions, PIII, microwave and ms-anneals. In *Ion Implantation Applications, Science and Technology*, ed. J. Ziegler, 12-1–12-68. Ion Implantation Technology Co., MD, 2010.

10. K. Tsukamoto, T. Kuroi, and Y. Kawasaki. Evolution of ion implantation technology and its contribution to semiconductor industry. In *Ion Implantation Technology 2010*, eds. J. Matsuo, M. Kase, T. Aoki, and T. Seki, 9–16. AIP Conference Proceedings 1321. American Institute of Physics, New York, 2010.

11. Y. Kawasaki, M. Ishibashi, M. Kitazawa, Y. Maruyama, S. Endo, T. Yamashita, and T. Kuori. Application of cluster boron implantation to pMOSFET's. In *Ion Implantation Technology 2010*, eds. J. Matsuo, M. Kase, T. Aoki, and T. Seki, 83–88. AIP Conference Proceedings 1321. American Institute of Physics, New York, 2010.

12. P. M. Fahey, P.B. Griffin, and J. D. Plummer. Point defects and dopant diffusion in silicon. *Rev. Modern Phys.*, 61, 289–384, 1989.

13. K. S. Jones and G. A. Rozgonyi. Extended defects from ion implantation and annealing. In *Rapid Thermal Processing Science and Technology*, ed. R. B. Fair, 123–168. Academic Press, Boston, 1993.

14. K. Goto, J. Matsuo, T. Y. Tada, T. Tanaka, Y. Momiyama, T. Sugii, and I. Yamada. A high performance 50 nm PMOSFET using decaborane ($B_{10}H_{14}$) ion implantation and 2-step activation annealing process. *IEDM Tech. Dig.*, 18.4.1–18.4.4, 1997.

15. D. Takeuchi, N. Shimada, J. Matsuo, and I. Yamada. Shallow junction formation by polyatomic cluster ion implantation. In *Proceedings of the 11th International Conference on Ion Implantation Technology—IIT '96*, vol. 1, issue 1, pp. 772–775. Institute of Electrical and Electronics Engineers, Piscataway, NJ, 1996.

16. T. Kusaba, N. Shimada, T. Aoki, J. Matsuo, I. Yamada, K. Goto, and T. Sugii. Boron diffusion in ultra low-energy (<1 keV/atom) decanorane ($B_{10}H_{14}$) ion implantation. In *1998 International Conference on Ion Implantation Technology Proceedings*, 1258–1261. Institute of Electrical and Electronics Engineers, Piscataway, NJ, 1999.

17. A. Agarwal, D. J. Eaglesham, H.-J. Gossmann, L. Pelaz, S. B. Herner, D. C. Jacobson, T. E. Haynes, Y. Erokhin, and R. Simonton. Boron-enhanced-diffusion of boron: The limiting factor for ultra-shallow junction. *IEDM Tech. Dig.*, 467–470, 1997.

18. Z. Insepov, T. Aoki, J. Matsuo, and I. Yamada. Computer simulation of annealing after cluster ion implantation. *Mater. Res. Soc. Proc.*, 532, 147–152, 1998.

19. D. Jacobson. Using boron cluster ion implantation to fabricate ultra-shallow junctions. In *Extended Abstracts of the 5th International Workshop on Junction Technology*, 27–30. Institute of Electrical and Electronics Engineers, Piscataway, NJ, 22005.

20. A. Renau. A better approach to molecular implantation. In *Extended Abstracts of the 7th International Workshop on Junction Technology*, 107–111. Institute of Electrical and Electronics Engineers, Piscataway, NJ, 2007.

8
Cluster Ion Beam–Assisted Deposition

Cluster ion beams can be utilized to enhance reactive growth of thin films at low substrate temperatures. The high density of transient energy produced by individual cluster impacts on a surface can significantly increase the rates of chemical reactions on the surface even at low temperatures. Surface smoothing and microstructure control take place during deposition of single-layer or multiple-layer thin films. High chemical reactivities resulting in good stoichiometric composition and effects that result in excellent surface morphology and in well-controlled film microstructure have been demonstrated by using O_2 gas cluster ion–assisted deposition. Representative examples are discussed below for Ta_2O_5 and diamondlike carbon (DLC) film depositions.

Obtaining direct experimental evidence of the high-temperature and high-pressure effects that occur during GCIB bombardment on solid surfaces has been a problem, even though such effects are consistently predicted by MD simulations (see, for example, Figure 2.19). One illustrative demonstration has resulted in the case of DLC films deposited by sublimation of C_{60} during simultaneous Ar-GCIB irradiation. As will be described in Section 8.2, the deposited DLC films have exhibited very significant increases of sp^3 hybridized carbon content in their atomic structures, and this has been attributed to the transient high-temperature and high-pressure conditions created by cluster ion impact.

8.1 Oxide Films

Figure 8.1 shows a diagram of a GCIB system designed and constructed by Epion Corporation U.S. specifically for assisted deposition of thin films [1, 2]. The equipment included a GCIB source, two electron beam evaporation sources, and a motor-driven rotating substrate holder. For depositing oxide films such as Ta_2O_5, Nb_2O_5, and SiO_2, appropriate oxide pellets were used as evaporation source materials. An O_2-GCIB gas cluster beam was produced by employing an O_2-He gas mixture (30% He). The presence of He in the gas mixture aided formation of the O_2 clusters by removing heat during expansion of the gas through the nozzle, but the He was itself not incorporated into the clusters. The average size of the O_2 clusters was in the range of 1000–3000 atoms/cluster. O_2 clusters were ionized by

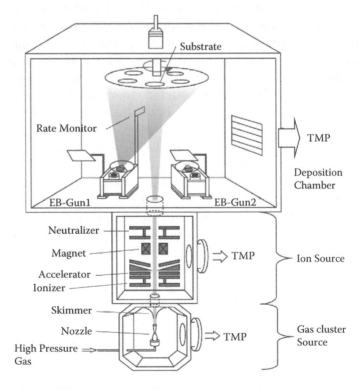

Figure 8.1 GCIB-assisted thin-film deposition system.

electron bombardment, and ionized clusters were then accelerated through potentials of up to 30 kV to target substrates at ground potential.

Figure 8.2 shows the average surface roughness of Ta_2O_5 films deposited without O_2-GCIB assist, with neutral O_2 cluster beam irradiation, and with O_2-GCIB irradiation at 7 keV. The O_2 neutral cluster condition means that the clusters were not ionized or accelerated [3]. The deposition chamber pressure was below 1×10^{-4} Torr, while the O_2 cluster beam was incident into the chamber. The film deposition rate and the thickness of the deposited Ta_2O_5 films were 0.1 nm/s and 180 nm, respectively. In the case of Ta_2O_5 films deposited without O_2-GCIB, the average surface roughness was 1.78 nm, and there were many bumps due to the columnar structure of the film. When the neutral O_2 cluster beam was used, the average surface roughness did not show significant reduction.

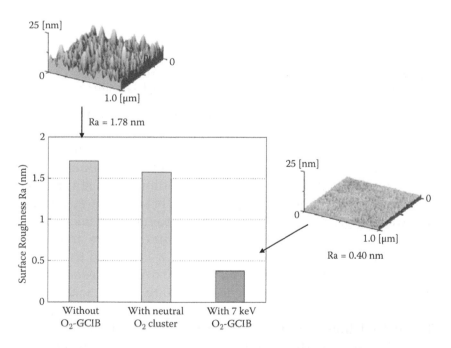

Figure 8.2 Average surface roughness of Ta_2O_5 films deposited without O_2-GCIB assist, with neutral O_2 cluster beam, and with 7 keV O_2-GCIB irradiation.

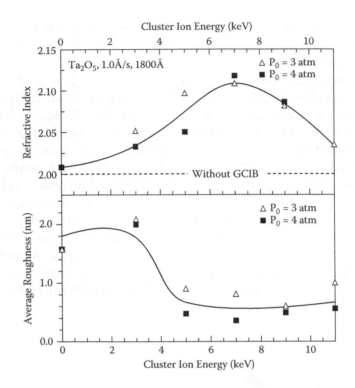

Figure 8.3 Acceleration energy dependence of refractive indexes and surface roughness of Ta_2O_5 films deposited with O_2-GCIB irradiations.

However, with O_2 cluster ion beam irradiation at an energy of 7 keV, the average surface roughness decreased to 0.4 nm.

Figure 8.3 shows the acceleration energy dependence of refractive index and surface roughness of Ta_2O_5 films deposited using simultaneous O_2-GCIB irradiation at stagnation pressures P_0 of 3 and 4 atm [3]. Refractive indexes measured at a wavelength of 633 nm are shown. Atomic force microscope (AFM) measurements of surface roughness were made on 1 μm squares. The thickness of the Ta_2O_5 films and the GCIB ion current density were 180 nm and 160 nA/cm², respectively. Acceleration energies were varied from 0 to 10 keV, with 0 keV representing the neutral O_2 cluster beam irradiation condition. The dotted line in Figure 8.3 represents the data for Ta_2O_5 films deposited without O_2-GCIB irradiations. With increasing acceleration energies, the refractive index increased and showed a peak at 7 keV. The Ta_2O_5 films deposited with

O_2-GCIB were transparent throughout the visible wavelength region. Surface roughness decreased to approximately 0.5 nm when cluster energy was increased to 5 keV, and it remained constant with further increase of the energy.

Figure 8.4 shows ion current density dependence of refractive indexes and surface roughness of Ta_2O_5 films deposited with O_2-GCIB assistance [3]. For comparison, the characteristics of SiO_2 films that were deposited using the same conditions are shown. The acceleration energy of O_2-GCIB was fixed at 7 keV. An ion current density of 0 represents the result with neutral O_2 cluster beams. In the case of neutral O_2 cluster beam–assisted deposition, the refractive index of the Ta_2O_5 was the same as that resulting from electron beam deposition without cluster beam assistance. As cluster ion current density was increased, the refractive index increased and reached a maximum value of 2.2 at ion current densities of 0.8 µA/cm²

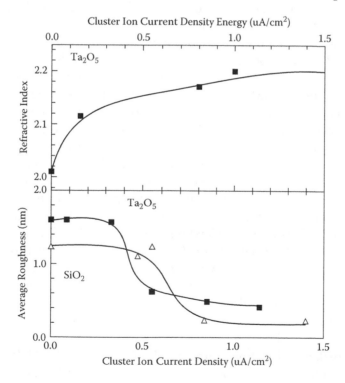

Figure 8.4 Ion current density dependence of refractive indexes and surface roughness of Ta_2O_5 and SiO_2 films.

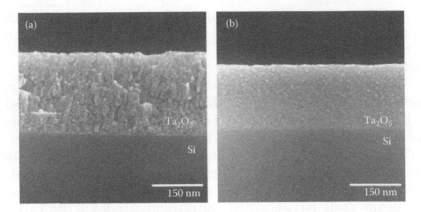

Figure 8.5 Cross-sectional SEM images of Ta_2O_5 films deposited (a) without O_2-GCIB and (b) with 7 keV O_2-GCIB.

and above. The surface roughness decreased from 1.6 nm at low ion current density to 0.6 nm, when ion current density reached 0.5 μA/cm². In the case of SiO_2, the surface roughness decreased when ion current density reached about 0.7 μA/cm².

GCIB irradiation was found to have a strong influence upon the microstructure of the Ta_2O_5 films [2]. Figure 8.5a shows a cross-sectional SEM image of a Ta_2O_5 film deposited without O_2-GCIB irradiation, and Figure 8.5b shows an image from a film deposited using O_2-GCIB irradiation at a cluster ion energy of 7 keV and an ion current density of 1.0 μA/cm². From these figures it can be seen that without O_2-GCIB irradiation, the film structure was coarse and columnar, but with 7 keV O_2-GCIB assistance, the film structure was very smooth and uniform. The composition of representative tantalum oxide films was measured by means of Rutherford backscattering [4]. The measurements showed that without O_2-GCIB assistance, the composition was $Ta_2O_{4.64}$, and with O_2-GCIB assistance, the composition was $Ta_2O_{4.98}$. Taking experimental error into account, this result shows that the tantalum oxide film deposited using O_2-GCIB assistance had been essentially fully oxidized to have stoichiometric composition.

In order to further examine the influence of GCIB assistance during film deposition, multiple layer film structures were made with alternating TiO_2 and SiO_2 layers deposited by e-beam evaporation of oxide pellets using O_2-GCIB assistance.

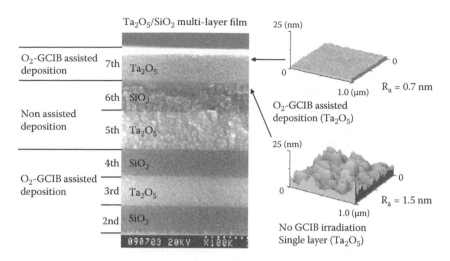

Figure 8.6 SEM cross-sectional image of Ta_2O_5/SiO_2 multilayer film and AFM images of individual layer surfaces.

Deposition rates were approximately 0.1 nm/s. For purposes of comparison, some of the e-beam evaporated film layers were deposited without O_2-GCIB assistance [5]. Figure 8.6 shows a cross-sectional SEM image of one example Ta_2O_5/SiO_2 multilayer film stack, and the figure also shows AFM images of the surfaces of the final two individual layers, one deposited with O_2-GCIB assistance and the other without. The lighter gray and dark layers correspond to the Ta_2O_5 and SiO_2, respectively. The second, third, fourth, and seventh layers from the bottom were formed with 7 keV O_2-GCIB assistance at a cluster ion current density of 1 $\mu A/cm^2$. The fifth and sixth layers were deposited without GCIB assistance. Each of the O_2-GCIB-assisted films can clearly be seen to be uniform with flat interfaces and without porous regions or columnar structures. In contrast, the layers produced without GCIB assistance exhibited large porous grains and rough interfaces.

8.2 Diamondlike Carbon Films

DLC films with a wide range of properties can be produced by many techniques, such as direct ion beam deposition, ion beam–assisted deposition, cathodic vacuum arc deposition, plasma deposition, pulsed laser deposition, and so forth [6–8]. The best deposition method for any particular DLC film application can be selected on the basis of the desired combination of film properties, such as hardness, stress, brittleness, smoothness, lubricity and adhesion. As an example of films with several excellent characteristics, those produced by vacuum arc deposition are reported to have very high density (3.1 g cm^{-3}), very high hardness (80 GPa), and high sp^3 diamond bond content (80%–88%) [8].

A highly stable low-energy ion-assisted deposition technique for producing high-quality DLC films has been achieved by combining carbon deposition from a C_{60} sublimation source with simultaneous Ar-GCIB bombardment [9]. A schematic diagram of the apparatus is shown in Figure 8.7.

Most forms of carbon, for example, graphite, have high thermal stability and are not convenient evaporation sources for pure carbon vapor, but C_{60} sublimes readily at temperatures of about 500°C and can be easily evaporated from a simple graphite crucible to deposit onto a substrate at room temperature. Concurrent bombardment by Ar cluster ions causes dissociation of the C_{60} molecules and produces a hard and very

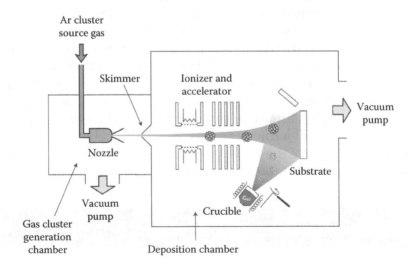

Figure 8.7 Schematic diagram of apparatus for DLC film deposition using C_{60} sublimation plus Ar-GCIB bombardment.

smooth DLC film. Because the DLC films produced by this approach do not contain hydrogen, they can be expected to remain stable at high temperatures.

For experimental studies, chamber vacuum pressures during deposition were below 3×10^{-5} Torr, C_{60} deposition rates were typically 15 nm/min, mean Ar cluster size was 1000 atoms, and the GCIB ion energy was between 3 and 9 keV. During initial investigations of the film formation behaviors, the ratio of the evaporation rate of C_{60} to the Ar cluster ion beam current density was studied. As is usual in all ion beam–assisted depositions, it is important to find a proper ratio of deposition rate to ion arrival rate. It was expected that when the C_{60}/Ar cluster ratio was high, mainly deposition would occur, but at low C_{60}/Ar cluster ratios, many physisorbed C_{60} molecules would be easily sputtered from the substrates. These were related to the energy density required to enhance phase transition from the sp^2 orbitals of C_{60} bonding to the sp^3 orbitals of diamond bonding.

Figure 8.8 shows DLC film growth rate versus the ratio R of the C_{60} molecule arrival rate relative to the Ar cluster ion

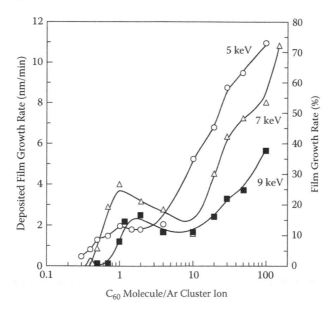

Figure 8.8 Dependence of DLC film growth rate on ratio R of C_{60} molecule arrival rate to Ar cluster ion arrival rate for ion energies of 5, 7, and 9 keV.

arrival rate for ion energies of 5, 7, and 9 keV [10]. The film growth rate represents the actual rate of film accumulation on the substrate as measured by a film thickness monitor. The Ar cluster ion arrival rate was obtained from the measured current density at the substrate. The second vertical axis on the right side of Figure 8.8 represents the DLC film deposition rates as percentages of the C_{60} film deposition rate measured without any cluster ion bombardment.

Figure 8.9 (I) shows film growth rate data from Figure 8.8 for the case of 7 keV cluster ions [10]. The rate of C_{60} film deposition without ion bombardment was 15 nm/min. In the region marked A on the figure, where the R ratio was less than 0.5, no film was deposited onto the substrates. In the region

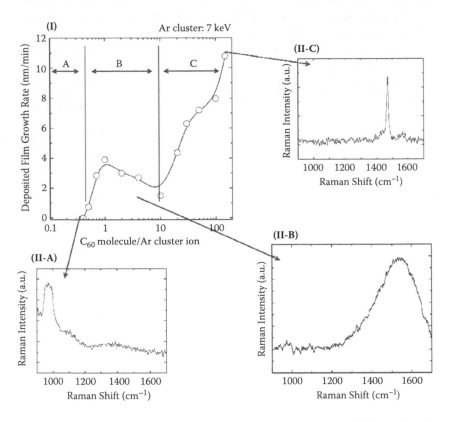

Figure 8.9 The dependence of DLC growth rate upon ratio of C_{60} molecule arrival rate to 7 keV Ar cluster ion arrival rate and example Raman spectra from representative films.

marked B ($0.5 < R < 10$), the film growth rate stayed between 2.0 and 3.8 nm/min. In region C ($R > 10$), the film growth rate increased monotonically with R, finally reaching a maximum of 75% of the rate without ion bombardment at R of 100. This result indicated that even when the C_{60} molecule arrival rate was 100 times higher than the Ar cluster ion arrival rate, a substantial amount of the C_{60} was still being sputtered from the surfaces by the ions.

As has been mentioned previously, the film properties should depend upon the ratio R. To confirm this, samples deposited under the conditions of each of the regions A, B, and C of Figure 8.9 (I) were studied using Raman spectroscopy. Figure 8.9 (II-A, II-B, and II-C) shows the Raman spectra from films deposited under the representative conditions of the A, B, and C regions. Figure 8.9 (II-A) is the typical Raman spectrum of a sample from region A, which shows only a silicon peak, indicating the presence of only the substrate itself, because depositing carbon had been removed by ion bombardment as quickly as it arrived. The Raman spectrum in Figure 8.9 (II-B) from a typical region B sample was similar to characteristic spectra from DLC films deposited using various other techniques, confirming the diamondlike nature of the region B films [6–8]. Figure 8.9 (II-C) shows a Raman spectrum from a sample deposited under the region C conditions. The spectrum shows a sharp peak related to C_{60} itself, indicating that the films deposited under region C conditions remained mainly as deposited C_{60}. Overall, these results indicated that DLC films could be deposited using R ratios between 0.5 and 10. It should be expected that the characteristics of DLC films deposited over this range of R ratios might vary appreciably.

Figure 8.10 shows AFM images of films that were deposited at different acceleration energies of 3, 5, and 7 keV [9]. Surface roughnesses of the films were (a) 0.3 nm at 3 keV, (b) 0.2 nm at 5 keV, and (c) 0.5 nm at 7 keV. The hardness of the DLC films was investigated using a nanoindentation method. Figure 8.11 shows measured hardness values versus acceleration energy for 0.3 µm thick films deposited using an R ratio of 2 [11]. The highest hardness resulted at an acceleration energy of 5 keV.

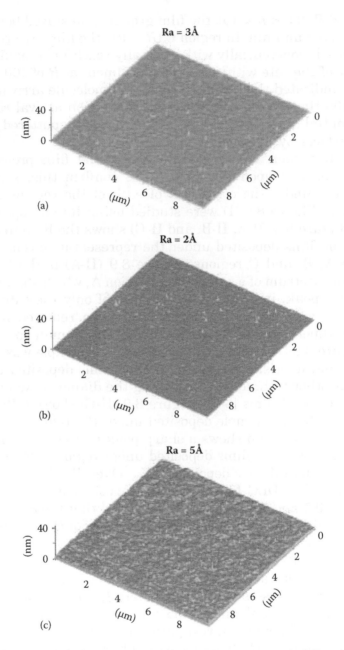

Figure 8.10 AFM images of the surfaces of films formed at different acceleration energies: (a) 3keV, (b) 5 keV, and (c) 7 keV.

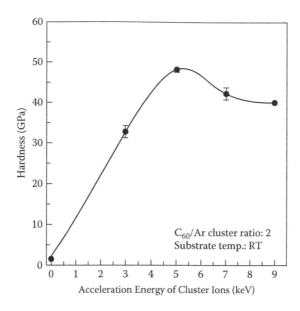

Figure 8.11 Acceleration energy dependence of hardness of the films deposited at $R = 2$ (C_{60}/Ar cluster ion).

Characteristic properties of carbon films are dependent upon their relative contents of sp^3 (diamond) and sp^2 (graphite) bonds. Milani and coworkers used near-edge x-ray absorption fine-structure (NEXAFS) spectroscopy for DLC evaluations [12]. They reported that NEXAFS spectroscopy was an efficient method for estimation of sp^2 content in amorphous carbon films. NEXAFS spectra obtained from films deposited using GCIB assistance have been reported [13]. Figure 8.12 shows example spectra obtained from various carbon materials, including a C_{60} film, glassy carbon, polycrystalline graphite, highly oriented pyrolytic graphite (HOPG), and a DLC film formed by C_{60} deposited with 5 keV Ar-GCIB bombardment. A spectrum from a polycrystalline diamond film deposited by chemical vapor deposition (CVD) is included for comparison [11].

Contents of sp^2 bonds in the various carbon materials shown in the NEXSFS spectra of Figure 8.12 can be compared. A pre-edge resonance at 285 eV is due to transitions from the C 1s orbital to the unoccupied π^* orbitals principally originating from sp^2 (C=C) sites, including the contribution of

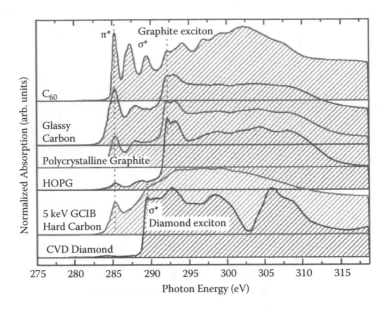

Figure 8.12 NEXAFS spectra from various carbon materials, including a C_{60} film, glassy carbon, polycrystalline graphite, HOPG, DLC film from C_{60} with 5 keV Ar-GCIB bombardment, and CVD diamond film.

sp (C≡C) sites if present. This peak is almost not visible in the spectrum of diamond, because diamond consists of only carbon atoms in the sp^3 (C–C) sites. Since the resonant peak π^* at 285 eV is the characteristic of unsaturated (sp^2) carbon bonds and a broad σ^* peak around 310 eV arises from a superposition of the signatures of many different bonding (sp^2 and sp^3) configurations in the matrix, the ratio of $sp^2/(sp^2 + sp^3)$ can be estimated [14–16]. It can be simply said that a decrease of sp^2 bonds increases sp^3 bonds.

The amount of sp^2 bonded carbon atoms can be extracted by the area of the resonance corresponding to $1s \rightarrow \pi^*$ transitions at 285 eV. By comparing this ratio with the ratio obtained in the same manner for a reference material, the sp^2 content relative to that of the reference material can be determined.

Figure 8.13 shows NEXAFS spectra from DLC films deposited by C_{60} with GCIB bombardment at 5, 7, and 9 keV with an R ratio of 2. For comparison, graphite and films deposited using ion plating, electron cyclotoron resonance (ECR) plasma, and RF plasma are also shown [17]. As was discussed

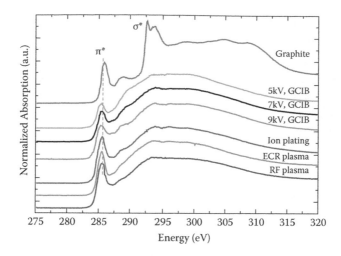

Figure 8.13 NEXAFS spectra of GCIB DLC films deposited at 5, 7, and 9 keV at $R = 2$. For comparison, films formed by ion-plating ECR plasma and RF plasma are also shown.

above, sp^2 ratios in all of the materials can be estimated relative to sp^2 content in DLC films deposited by RF plasma by comparing the areas of the resonances corresponding to $1s \rightarrow \pi^*$ transitions at 285 eV. Figure 8.14 shows the sp^2 contents determined in this manner presented as percentages of the sp^2 content of RF plasma-deposited DLC. It should be noted that in the case of the GCIB-assisted DLC films, the sp^2 contents increased as the cluster ion energy increased from 5 to 9 keV [18]. It is reasonable to expect the most diamondlike behavior from the materials with the highest sp^3 bond contents, that is, those with the lowest sp^2 contents.

For DLC films produced using Ar-GCIB bombardment at 5, 7, and 9 keV, and for a few example films produced by other deposition methods, hardness values measured by nanoindentation and relative sp^2 contents estimated from NEXAFS spectra are listed in Table 8.1 [14, 19]. The estimated relative sp^2 contents of the DLC films formed by the GCIB method were significantly lower than those of the DLC films formed by the other methods. The hardness values listed in Table 8.1 for the GCIB-assisted DLC films were in the range of 40–50 GPa, while the values for the DLC films formed by the other methods were closer to 20 GPa. From inspection of this table it can

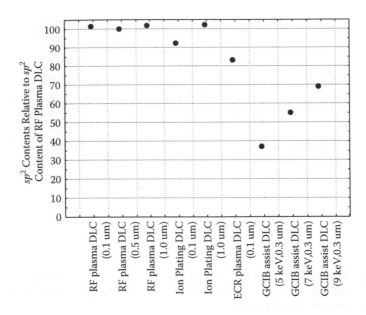

Figure 8.14 Estimated sp^2 contents of DLC films as percentages of the sp^2 content in RF plasma-deposited DLC.

TABLE 8.1
Relative sp^2 Contents and Hardness Values

Method (Acceleration Value)	Relative sp^2 Content	Hardness Value (GPa)
GCIB (5 kV)	36	48.3
GCIB (7 kV)	54	42.3
GCIB (9 kV)	68	40.0
RF plasma	100	18
ECR plasma	82	23
Ion plating	91	26

be recognized that there is clearly a correlation between relative sp^2 contents and the hardness of the films.

Conductive boron-doped DLC films with high hardness have been produced by GCIB-assisted deposition [20, 21]. Using the equipment shown in Figure 8.7, vapor of a boron hydride molecular compound, usually decaborane ($B_{10}H_{14}$), was introduced into the ionizer region. The boron-containing molecules were ionized by the same electron bombardment used to ionize the Ar cluster beam, and the resulting ions were then accelerated through the same potential as the cluster ions so as to also bombard the growing DLC film. Resulting films showed good conductivity with reported resistivities of approximately 3×10^{-3} ohm·cm. Hardness of the films was measured to be 40 GPa, which is approximately the same as the values measured on the undoped GCIB films discussed above. Figure 8.15 shows the results of a test to determine the influence of temperature upon the electrical resistivity of the boron-doped DLC films. The film resistivity remained essentially unchanged as a result of 8 h long exposures in air at temperatures ranging from 25°C to 400°C. At temperatures above 400°C, film loss occurred due to oxidation.

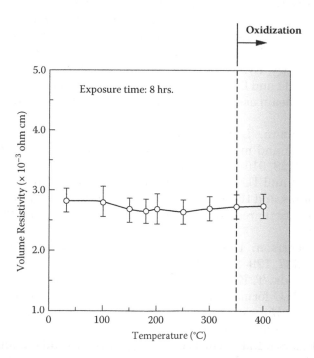

Figure 8.15 The influence of thermal treatment upon electrical resistivity of boron-doped DLC films.

References

1. Y. Fujiwara, N. Toyoda, and I. Yamada. Development of high current gas cluster ion beam assisted deposition system. In *Extended Abstracts of the 4th Workshop on Cluster Ion Beam and Advanced Quantum Beam Process Technology*, Tokyo, September 2003, pp. 98–100.

2. T. Nose, S. Inoue, N. Toyoda, K. Mochiji, T. Mitamura, and I. Yamada. Development of pure O_2 cluster ion beam assisted deposition system. *Nucl. Instrum. Methods Phys. Res. B*, 241, 626–629, 2005.

3. N. Toyoda and I. Yamada. High quality optical thin film formation with low energy gas cluster ion beam irradiation. In *Ion Implantation Technology 2002*, eds. B. brown, T. L. Alford, M. Nastasi, and M. C. Vella, 701–704. Institute of Electrical and Electronics Engineers, Piscataway, NJ, 2002.

4. Y. Fujiwara, N. Toyoda, and I. Yamada. Reduction of surface roughness by Ta_2O_5 film formation with O_2 cluster ion assisted deposition. In *3rd Workshop on Cluster Ion Beam Process Technology*, Kyoto, Japan, 2002, pp. 155–160.

5. N. Toyoda and I. Yamada. Optical thin film formation by oxygen cluster beam assisted deposition. *Appl. Surf. Sci.*, 226, 231–236, 2004.

6. J. Ullmann. Low energy ion assisted carbon film growth: Methods and mechanisms. *Nucl. Instrum. Methods Phys. Res. B*, 127/128, 910–917, 1997.

7. H. Tsai and D. B. Bogy. Characterization of diamondlike carbon films and their application as overcoats on thin film media for magnetic recording. *J. Vac. Sci. Technol. A*, 5, 3287–3312, 1987.

8. J. Robertson. Diamond-like amorphous carbon. *Mater. Sci. Eng.*, R37, 129–281, 2002.

9. I. Yamada, T. Kitagwa, J. Matsuo, and A. Kirkpatrick. Ultrahard DLC formation by gas cluster ion beam assisted deposition. In *Mass and Charge Transport in Inorganic Materials: Fundamentals to Devices*, eds. P. Vincenzini and A. Buscaglia, Techna Srl. Lido di Jesolo, Venice, Italy, 957–964. 2000.

10. T. Kitagawa, I. Yamada, N. Toyoda, H. Tsubakino, J. Matsuo, G. H. Takaoka, and A. Kirkpatrick. Hard DLC film formation by gas cluster ion beam assisted deposition. *Nucl. Instrum. Methods Phys. Res. B*, 201, 405–412, 2003.

11. T. Kitagawa. Super hard carbon coating with gas cluster ion beam assisted deposition. PhD thesis, University of Hyogo, 2003.
12. C. Lenardi, P. Piseri, V. Briois, C. E. Bottani, A. Li Bassi, and P. Milani. Near-edge x-ray absorption fine structure and Raman characterization of amorphous and nanostructured carbon films. *J. Appl. Phys.*, 85, 7159–7167, 1999.
13. T. Kitgawa, K. Miyauchi, N. Toyoda, K. Kanda, S. Matsui, H. Tsubakino, J. Matsuo, and I. Yamada. NEXAFS study of DLC films deposited with Ar cluster ion and Ar⁺ bombardment. In *2002 14th International Conference on Ion Implantation Technology Proceedings*, Taos, NM, September 22–27, 2002, pp. 587–590.
14. K. Kanda, T. Kitagawa, Y. Shimizugawa, Y. Haruyama, S. MatsuiI, M. Tetasawa, H. Tsubakino, I. Yamada, T. Gejo, and M. Kamada. Characterization of hard diamond-like carbon films formed by Ar gas cluster ion beam-assisted fullerene deposition. *Jpn. J. Appl. Phys.*, 41, 4295–4298, 2002.
15. M. Jaouen, G. Tourillon, J. Delafond, N. Junqua, and G. Hug. A NEXAFS characterization of ion-beam-assisted carbon-sputtered thin films. *Diamond Related Mater.*, 4, 200–206, 1995.
16. A. Saikubo, N. Yamada, K. Kanda, S. Matsui, T. Suzuki, K. Niihara, and H. Saitoh. Comprehensive classification of DLC films formed by various methods using NEXAFS measurement. *Diamond Related Mater.*, 17, 1743–1745, 2008.
17. K. Miyauchi. High hard DLC film deposition by gas cluster ion beams. Graduation thesis, Himeji Institute of Technology, 2002.
18. T. Kitagawa. Evaluation of DLC film formed by GCIB assisted deposition. Internal report. Himeji Institute of Technology, January 16, 2001.
19. F. Ohtani. Present status of gas cluster ion beam assisted deposition technique. In *3rd Workshop on Cluster Ion Beam Process Technology*, Kyoto, Japan, 2002, pp. 83–89.
20. T. Kitagawa. Novel carbonaceous film with high electrical conductivity and super high hardness for semiconductor test probes, Presented at 2011 IEEE SW Test Workshop, Session 2, San Diego, CA, June 2011. http://www.swtest.org/swtw_library/2011proc/swtw2011.html.
21. T. Kitagawa and S. Nomura. Conductive hard carbon film, and film forming method therefore. Patent WO2012073869 A1, filed November 28, 2011.

9

Applications

As characteristic dimensions of semiconductor, optical, magnetic, and mechanical devices on IC chips have progressed to below submicron levels (<100 nm), fabrication processes able to offer nanoscale accuracy have become increasingly important. Ion beam techniques have traditionally provided precise controllability of ion species, energy, and dose, but in order to satisfy new nanoscale processing requirements, very low-energy ion beam processes have become necessary. However, it is in principle very difficult to attain atomic ion beams that have both very low ion energies and, at the same time, high ion currents. Gas cluster ion beam techniques provide solutions to this inherent problem that faces conventional ion beam approaches. Discussed in this chapter are some of the original representative examples of gas cluster ion beam (GCIB) characteristics employed for nanoscale applications such as precise etching and smoothing for optical and magnetic devices, very low-energy implantation for ultrashallow junction formation, low-energy etching for x-ray photoelectron spectroscopy (XPS), and high-yield secondary ion production for secondary ion mass spectroscopy (SIMS), and for molecular response enhancement of biomedical implant materials.

9.1 Nanoscale Etching and Smoothing for Optical and Magnetic Devices

The very low-energy irradiation effect that is characteristic of GCIB can be employed to substantially reduce surface defects on almost any highly polished material. Chemical mechanical polishing (CMP) techniques used for preparation of optical surfaces can result in numerical surface roughness values of only a few angstroms, but the surfaces typically still contain high densities of shallow scratches and other defects. These flaws, which cannot be corrected by additional processing using conventional ion beams or electron beams, can often be eliminated easily by GCIB bombardment. A representative example is shown in Figure 9.1 for the case of the surface of a lens made of CaF_2, a relatively soft optical material upon which it is normally exceptionally difficult to eliminate scratches caused by mechanical polishing action. The atomic force microscope (AFM) images before and after polishing by Ar-GCIB clearly illustrate the characteristic smoothing behavior of GCIB [1].

On some optical materials, conventional polishing methods can leave residual asperities, small localized projections, which can be detrimental for many applications. GCIB bombardment has been shown to be highly effective for removing surface asperities. This has been found to have particular merit for a range of photonic applications. Surface roughness of glass substrates for narrow-bandpass dense-wavelength division-multiplexing (DWDM) filters used in optical telecommunications is an important example. After polishing to the limits achievable by CMP, the glass substrate surfaces are left with large numbers of extremely small local asperities having heights of the order of tens of angstroms. During subsequent deposition of more than 100 film layers to form the optical filter stack, the asperities on the substrate surface become replicated and amplified through the film stack, to cause scattering, significant performance degradation, and substantial production yield losses. Ar-GCIB smoothing of the glass substrate to eliminate surface asperities prior to deposition of the film layers has been found to prevent development of scattering centers, has resulted in optimization of the bandpass behavior, and has increased production yields. Figure 9.2 shows example AFM images of glass substrate surfaces before and after Ar-GCIB bombardment. GCIB processing can

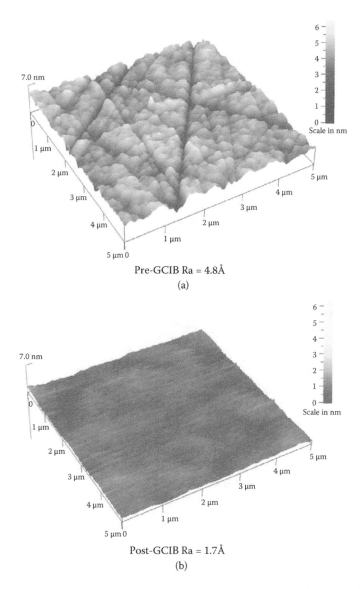

Pre-GCIB Ra = 4.8Å

(a)

Post-GCIB Ra = 1.7Å

(b)

Figure 9.1 AFM images of CaF₂ optical lens surfaces before and after GCIB processing.

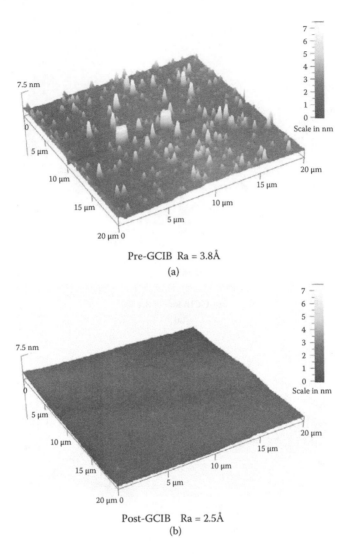

Figure 9.2 Asperity removal from glass substrates for DWDM filters.

be seen to have produced moderate reduction of the surface roughness value while eliminating asperities, so as to enormously improve the quality of the surface as a substrate for a complex film stack [1]. Similar advantages can be expected for other multilayer interference coatings and for critical surfaces on many aspheric elements, x-ray mirrors, extreme ultraviolet lithography (EUV) mirrors and masks, and so forth.

GCIB etching is often employed in a mode known as location-specific processing (LSP), which allows precisely controlled nonuniform removal of material from surfaces in order to correct surface uniformity errors or performance errors of device structures on the surfaces. GCIB production processors use mechanical scanning of the process targets through a stationary beam. In an LSP process, the computer that controls the GCIB system is provided with a map of the thickness of material to be removed from every position on the target surface, and the speed at which the target is moved past the beam is then continuously varied so as to produce the desired profile. Careful calibration of the removal rate produced by the GCIB in the material of interest in combination with an adequate technique for creating the planned removal map allows accurate surface profiles to be produced. Figure 9.3 shows an example of the result of an LSP process on an initially nonuniform SiO_2 film on a 200 mm silicon wafer. Initially, the film thickness varied by ±87 angstroms over the surface. After LSP removal of approximately 1800 angstroms from the surface, the oxide thickness variation was reduced to ±3 angstroms [1].

The GCIB-LSP technique has also been used for nondamaging thinning and uniformity improvement of active silicon layers on silicon-on-insulator (SOI) wafers. Figure 9.4 shows a SOI wafer example in which a Si surface layer of an average thickness of 145 nm with a standard deviation of 0.85 nm was reduced to a 50 nm thickness and standard deviation of 0.4 nm. The processing was conducted at 20 kV using 5% CF_4 in 95% Ar, followed by a 3 kV pure O_2-GCIB exposure [2].

The GCIB-LSP technology is presently employed in production of film bulk acoustic resonator (FBAR) devices for cell phones and GPS devices [3]. Operating frequencies of FBAR filter devices depend very critically upon thickness of a surface layer of a piezoelectric material such as AlN, and adequate frequency uniformity across individual wafers and from wafer to wafer is difficult to achieve. In order to increase production yields, LSP is used to remove precisely controlled amounts of surface material in patterns that are adjusted for each individual wafer. Figure 9.5 shows a reduction of width of frequency distribution by factor of 5 over a typical 200 mm substrate. It has been reported that piezoelectric quality of

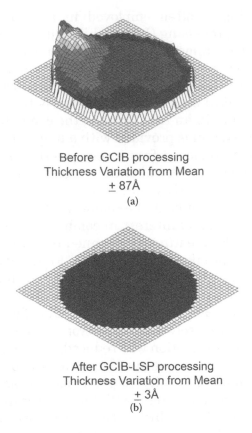

Before GCIB processing
Thickness Variation from Mean
± 87Å
(a)

After GCIB-LSP processing
Thickness Variation from Mean
± 3Å
(b)

Figure 9.3 Location-specific GCIB processing of SiO$_2$ film.

the highly textured, polycrystalline, c-axis-oriented columnar surface film is not affected by the GCIB exposure.

Etching and smoothing of very hard materials can be accomplished by GCIB processing. Silicon carbide (SiC) is an example of such a material. Standard mechanical polishing of SiC is typically done using a series of diamond-based slurries, where the abrasive size is successively reduced until eventually a submicron slurry is employed to produce a final result. Residual scratches that are typically left by the mechanical polishing can often be problem issues relative to applications. GCIB has been found to be effective in eliminating such scratches. Figure 9.6 shows an example of AFM images of the surface of a 6 in. diameter polycrystalline SiC wafer (a) as polished by

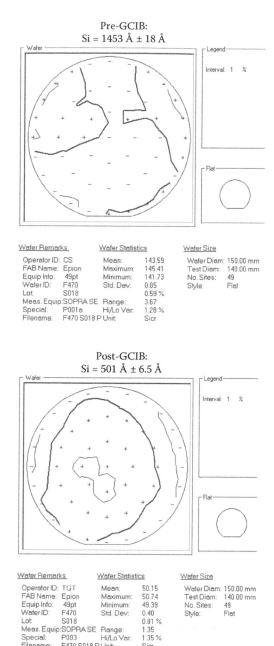

Figure 9.4 SOI thinning by GCIB processing.

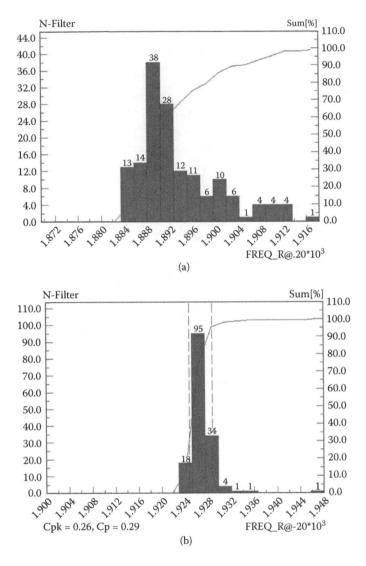

Figure 9.5 Frequency distribution of FBAR (a) before and (b) after GCIB etching.

CMP, (b) after 20 kV Ar-GCIB irradiation to a dose of 1×10^{16} ions/cm^2, and (c) after additional 10 kV O$_2$-GCIB irradiation to a dose of 1×10^{15} ions/cm^2. The initial CMP-polished surface was very flat, with 0.3 nm roughness, but many scratches were present. After the Ar-GCIB irradiation, the scratches

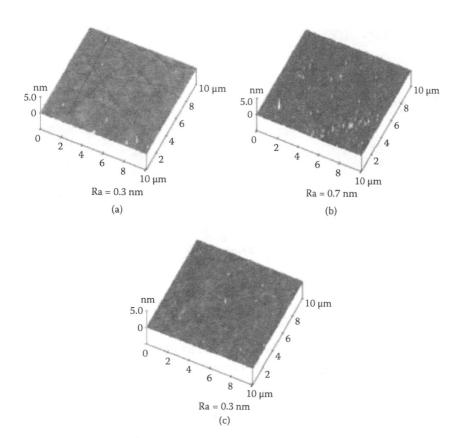

Figure 9.6 AFM image of poly-SiC surface (a) before irradiation, (b) after Ar-GCIB irradiation, and (c) after O_2-GCIB irradiation in addition to the Ar-GCIB irradiation.

were gone, but small asperities appeared and surface roughness increased to 0.7 nm. After the O_2-GCIB irradiation step, the number of asperities was reduced, average surface roughness decreased to 0.3 nm, which was the same as the initial roughness, and no grains were observed [4].

Surface smoothing of the sidewalls of high-aspect-ratio Si pillar structures for photonic crystal DWDM devices has been studied. For these types of tall pillar structure formations, successive etching and deposition by inductively coupled plasma-reactive ion etching (ICP-RIE) are usually applied. The processing typically results in excessively rough sidewalls. The silicon pillar structure shown in Figure 9.7 was irradiated

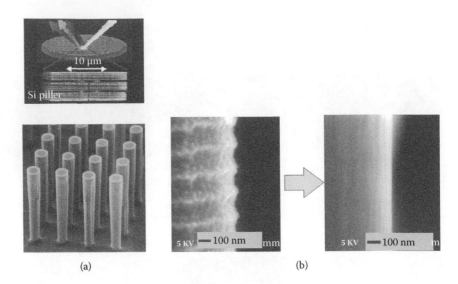

(a)   (b)

Figure 9.7 GCIB application for photonic devices. (a) Si pillar-type photonic device. (b) Surface smoothing of the pillar.

by SF_6 gas cluster ions at an incident angle of 83° from the surface normal to the sidewall. The acceleration voltage and the ion dose of the SF_6 clusters were 30 kV and 1.5×10^{15} ions/cm^2, respectively. Figure 9.7 shows (a) the structure of the photonic device and (b) the surface pre- and post-GCIB treatments. The surface protuberances were selectively removed after etching a depth of only 50 nm. It has been suggested that if ultra-smooth sidewalls with an average roughness of 0.3 nm can be fabricated, then DWDM device transmission losses can be reduced to levels lower than those presently resulting in photonic crystal structures [5, 6].

Surface smoothing of laser-crystallized polycrystalline silicon thin films has been investigated. A laser crystallization technique is a useful approach for obtaining high-quality poly-Si films. The laser process, however, results in formation of many hillocks near the grain boundaries of the poly-Si films. This effect makes it difficult to realize high-performance poly-Si thin-film transistors (TFTs). Under typical conditions, the thickness of the laser-crystallized poly-Si film is about 50 nm, and the created hillock height is almost the same as the film thickness. The ability to accomplish surface smoothing without thinning the film and without causing damage to the Si is an important

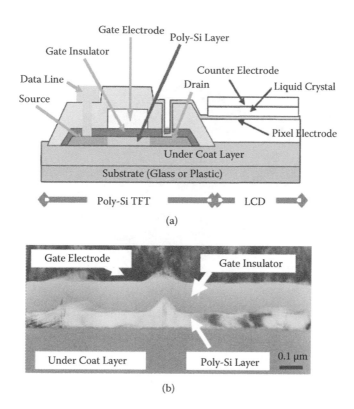

Figure 9.8 (a) Schematic of poly-Si TFT structure. (b) SEM image of the layers near the gate electrode.

factor. Figure 9.8 shows (a) the structure of a poly-Si TFT and (b) a scanning electron microscope (SEM) image of the layers of poly-Si, gate insulator, and gate electrode films after laser annealing. The hillock formation in the Si layer can be seen [6].

Figure 9.9 shows example AFM images from a poly-Si film surface (a) after laser crystallization and (b) after Ar-GCIB irradiation at 20 kV and 1×10^{16} ions/cm^2 [7]. By selecting cluster source materials, irradiation energy, combinations of different source materials, and glancing-angle irradiation processing, surface smoothing by GCIB has effectively eliminated the hillocks on poly-Si films. Damage caused by GCIB irradiation has also been investigated. Capacitance-voltage (C-V) measurements of metal-insulator-semiconductor (MIS) capacitors have shown that damage generated by GCIB irradiation can be suppressed by using glancing-angle bombardment.

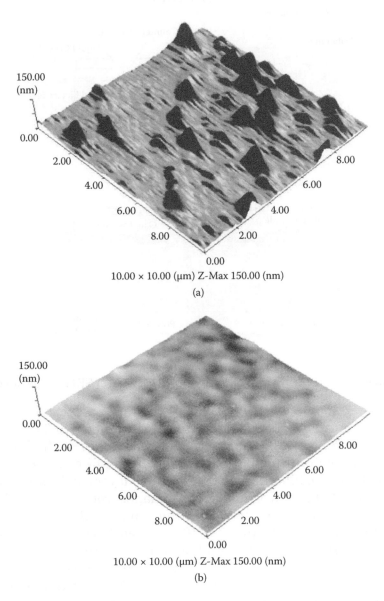

Figure 9.9 AFM images of a poly-Si film (a) after laser crystallization and (b) after GCIB irradiation.

Using SF_6-GCIB clusters of mean size 3000 molecules at an acceleration voltage of 20 kV, flat smooth poly-Si surfaces have been achieved while removing less than 2 nm of material [8].

GCIB processing of magnetic materials under conditions that result in extremely low damage has been studied. In order to increase areal density in hard disk drives (HDDs), it is necessary to decrease the flying height of the magnetic recording head over the disk. In order to do this, smoothing of the magnetic material surface can be extremely useful, because roughness of the surface limits the flying height and decreases the head sensitivity. However, it is important that the magnetic material properties should not be altered. Surface damage caused by GCIB irradiation has been evaluated by film strain measurements. Figure 9.10 shows (a) the structure of a HDD sensor and (b) the strain distribution within example PtMn films caused by GCIB and by Ar ion beam irradiation.

Film strain was measured by using grazing incidence synchrotron radiation x-ray diffractometry (XRD) at a wavelength of 0.15418 nm [9]. In this measurement, lattice parameters of face-centered cubic (fcc) (111) spacing in the in-plane (d_{ip}) (perpendicular to the surface) and out-of-plane (d_{op}) (parallel to the surface) directions were measured as functions of x-ray penetration depth. GCIB irradiation was done at 20 kV (ion dose: 3×10^{15} ions/cm^2) from an incident angle of 80°. For comparison, Ar-atomic ion beam irradiation was made at 100 V, also at an incident angle of 80°. The film strain was defined as $\{[(d_{op}) - (d_{ip})]/(d_{ip})\} \times 100$ (%). Strain depth distributions for as-deposited, Ar ion beam–irradiated, and GCIB-irradiated films are shown in Figure 9.10b. The as-deposited PtMn films showed compressive and tensile stresses in the bulk and surface regions, respectively. The films irradiated by GCIB and Ar ion beam exhibited compressive stress throughout the films. Note that etched depths by GCIB irradiation and Ar ion beam were 6.3 and 2.2 nm, respectively. The strain profile of the GCIB-irradiated film was similar to that of the as-deposited film. In contrast, the strain distribution of the Ar ion beam-irradiated film shifted to higher values throughout the film thickness. Surface smoothness and damaged layer thickness of a 50 nm thick PtMn film deposited on a Si substrate

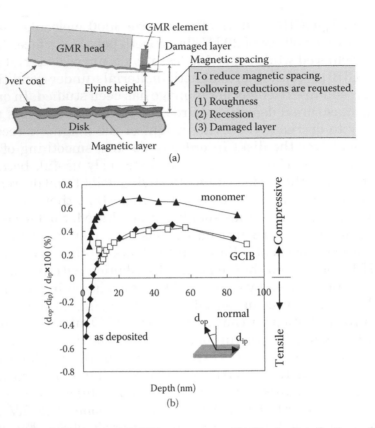

Figure 9.10 (a) Structure of HDD sensor area. (b) Strain distributions of as-deposited, GCIB, and monomer ion beam–irradiated PtMn films obtained from grazing incidence XRD as a function of x-ray penetration depth.

were also examined using measurement by AFM, SIMS, and x-ray reflection. The surface roughness Ra was 1.2 nm, and the damaged layer thickness was less than 1.5 nm

Among approaches to realizing higher HDD areal record-ing densities are discrete track media (DTM) and bit-patterned media (BPM) concepts in which discrete areas of magnetic material are employed. A smooth plane surface is required under the flyer head, and work has been conducted to dem-onstrate that GCIB can be used to accomplish the necessary structure [10]. For GCIB process evaluations, test substrates had line-and-space magnetic material patterns overcoated by continuous layers of nonmagnetic TiCr. GCIB was used to planarize and smooth the TiCr. Figure 9.11a shows an AFM

(a) Before irradiation   (b) AFM image after GCIB   (c) MFM image after GCIB

(PV = 21.5 nm, RMS = 8.4 nm)   (PV = 1.8 nm, RMS = 0.7 nm)

Figure 9.11 AFM and MFM images of pattern before and after GCIB planarization.

image of an as-deposited TiCr surface that has the patterned contour of the magnetic material below. Figure 9.11b is an AFM image of the same area showing the smooth plane TiCr surface following irradiation at 20 kV with Ar-GCIB to an ion dose of 1×10^{16} ions/cm^2 and N$_2$-GCIB to an ion dose of 1×10^{15} ions/cm^2. Root mean square (RMS) roughness of the TiCr surface was improved from the initial value of 8.4 nm to 0.7 nm. Figure 9.11c is a magnetic force microscopy (MFM) image of the same area that shows the magnetic pattern still intact beneath the processed TiCr film. The MFM image clearly shows that GCIB planarization was confirmed and the discrete track pattern of the magnetic layer was preserved.

9.2 Highly Anisotropic Etching Processing

Reactive gas cluster beams produced by supersonic nozzles have been employed for highly anisotropic etching processing. Figure 9.12 shows a schematic diagram of the equipment that consists essentially of a simple nozzle in vacuum [11, 12]. The cluster beam is produced via a supersonic nozzle, the same as is used for GCIB equipment, but in this case the beam is not ionized.

For a demonstration, ClF$_3$ gas diluted to several percent by Ar gas was used. The source gas pressure to the nozzle was

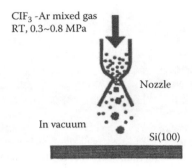

Figure 9.12 Schematic of reactive cluster gas etching tool. For Si etching, a ClF_3-Ar gas mixture is introduced at inlet pressure of 0.3–0.8 GPa.

0.3–0.85 MPa. The energy of an as-generated neutral cluster was estimated to be less than 1 eV/atom (or eV/molecule). Since the clusters are neutral, irradiation can be made without the influence of beam divergence, which is usually present in ion beams due to space charge effects. Because of these characteristics, anisotropic, highly selective, very precise, and damage-free etching can be achieved.

Figure 9.13 shows the etching rate of Si(100) irradiated by ClF_3-Ar neutral cluster beams produced at different source

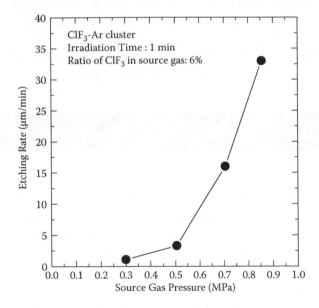

Figure 9.13 Si(100) etching rate versus source gas pressure.

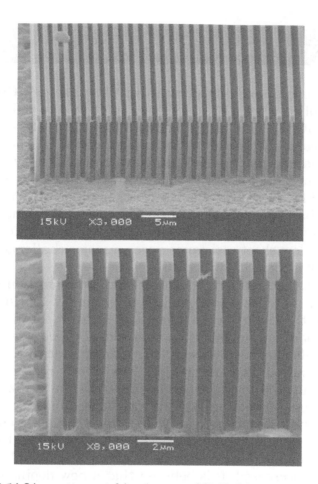

Figure 9.14 Line pattern etching images of Si. Etching was performed by irradiation of CF_3 (6%)/Ar for 15 min.

gas pressures. The pressure in the process chamber was kept at less than 10 Pa during the irradiations. For this demonstration, a Si(100) substrate was placed at a distance of 13 mm from the nozzle exit. The etching rate increased nonlinearly with source gas pressure and exceeded 30 μm/min at 0.85 MPa. The threshold pressure for etching was about 0.3 MPa. Figure 9.14 shows a SEM image of a Si substrate that was etched behind a 1.2 μm thick mask having a 0.75 μm line width. Under these conditions, the etching ratios of Si/SiO_2 and Si/resist were higher than 1000. Highly precise anisotropic etching with an aspect ratio of 7 was realized.

9.3 Semiconductor Device Fabrication

Very low-energy and high-current characteristics of GCIB have been applied for metal oxide semiconductor field-effect transistor (MOSFET) shallow junction formation. It is becoming increasingly difficult to accomplish MOSFET source and drain shallow junction formation with satisfactory abruptness to meet requirements for future technology nodes [13]. Figure 9.15 shows the schematic cross section of a MOSFET device where junction depth is x_j and physical gate length is L_g [14]. The formation of sufficiently ultrashallow junctions with adequate low sheet resistance poses a major challenge in the scaling of advanced devices [15].

It has been realized that ever-shallower junction depths cannot in practice be accomplished by simply continuing to decrease the energy of the dopant ion implant. In the case of conventional ion implantation technology, which uses atomic or molecular ion beams, beam current degradation and limits at the required ultralow energies can greatly reduce process throughput and increase equipment costs. In the implantation process, dopant self-sputtering, deposition, and etching are also limiting factors relative to lowering implant energies. The transient enhanced diffusion (TED) effect and dopant deactivation also become serious problems as ion implantation energy is decreased. It is believed that a new doping concept

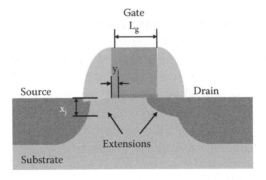

Figure 9.15 Schematic cross section of a MOS transistor, defining vertical junction depth x_j and physical gate length L_g.

is needed in order to form ultrashallow junctions with lower sheet resistances and lower densities of crystal defects [16, 17].

As ultrashallow junction depths become further reduced, attainable reductions of gate length L_g must also be considered. The reduction of the gate length to dimensions comparable to depths of the drain and source junctions affects the device operational speed and other device characteristics, such as threshold voltage (V_{th}), off-state leakage current (I_{off}), drive currents (I_{on}), and so forth. These are called short-channel effects (SCEs). As gate length shrinks, the voltage that can be applied to the gate must be reduced to maintain reliability, and the threshold voltage has to be reduced as well. As threshold voltage is reduced, the transistor cannot be switched from complete turn-off to complete turn-on with the limited voltage swing available. V_{th} roll-off is one of the most serious consequences of SCEs, and this characteristic determines the minimum L_g that will be acceptable. GCIB and polyatomic ion implantations for ultrashallow junction formation have been studied in terms of SCEs.

9.3.1 GCIB Infusion Doping

GCIB infusion doping of boron for ultrashallow junctions was reported for the first time in 2004 [18]. Ultrashallow junctions of 12 nm depth were produced using a mixture of B_2H_6 and Ar gas GCIB at 5 kV. Dopant profiles measured by SIMS exhibited extreme abruptness of <2.5 nm/decade for a 12 nm shallow junction and showed no evidence of channeling behavior. Infusion doping showed a log to the 1/3 power relationship between energy and junction depth in contrast to the linear fit, which is traditionally observed with conventional ion implantation. Boron surface doping concentrations of 1–2×10^{22}/cm^3 for 2×10^{16}/cm^2 doses were achieved.

In the GCIB infusion doping process, a linear relationship was found to exist between the GCIB dose and the retained B dose in Si from low GCIB doses of less than 1×10^{15} ions/cm^2 through very high doses of more than 1×10^{17} ions/cm^2. Due to self-sputtering limitations, a similar relationship does not exist with conventional ion implantation by atomic or molecular ion beams. The linear relationship in the case of GCIB infusion doping has been observed to hold for cluster implantation

up to 60 kV [19]. As has been described previously, the competing effects among implantation, deposition, and etching that
become problematic at low implant energies and high doses
in the case of conventional atomic or molecular ion beams
can clearly be eliminated in the infusion doping process. Low
sheet resistances have been achieved for ultrashallow junctions produced using high GCIB infusion doses. One reported
example described an electrical junction depth of 5 nm with a
sheet resistance of 350 ohms/square after a 550°C activation
anneal [18].

Sub-50 nm positive metal oxide semiconductor field-effect
transistors (p-MOSFETs) fabricated using GCIB infusion doping were reported in 2005 [20]. Infusion doping of B was conducted at 5 kV to a dose of 1×10^{15} ions/cm^2. For comparison,
devices were also fabricated using conventional low-energy ion
implantation doping at 0.2 kV, also to 1×10^{15} ions/cm^2 dose.
After the doping processes, spike annealing at 1050°C was
used for activation. Figure 9.16 shows threshold voltage characteristics of the p-MOSFETs as functions of gate length L_g.
The use of infusion doping improved SCE by about 20 nm relative to the results with low-energy conventional implantation,
and superior roll-off characteristics were obtained for gate
lengths of ≤50 nm. Figure 9.17 shows the I_{off}-I_{on} characteristics

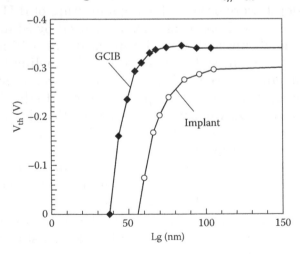

Figure 9.16 V_{th}-L_g characteristics of first p-MOSFETs by infusion doping
at 5 kV, compared with conventional ion implantation at 0.2 kV.

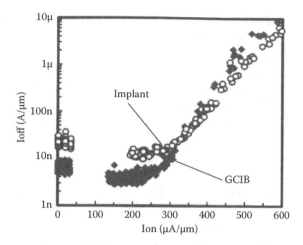

Figure 9.17 I_{off}-I_{on} characteristics of first p-MOSFETs by infusion doping, compared with those by conventional ion implantation.

of the devices. Reduction of off-state leakage current compared with that by conventional low-energy ion implantation was obtained. On-current as high as ~400 μA/μm was obtained at 100 nA/μm off-current. The results indicated that using a simple replacement of low-energy boron implantation by GCIB infusion doping, about a 20 nm improvement in the short-channel effect and almost the same current drivability were obtained for the p-MOSFETs.

In 2007, by improving the CMOS fabrication processes, especially by photoresist removal, sub-30 nm p-MOSFETs were reported [21]. These devices showed leakage currents I_{off} of 1 nA/μm and I_{on} of ~300 μA/μm. The short channel effect improved by about 10 nm compared to that resulting with conventional ion implantation. The results demonstrated that the GCIB doping enabled prospects for shallower B doping profiles.

9.3.2 Polyatomic Cluster Ion Doping

In 1996, the first p-MOSFETs fabricated using decaborane ($B_{10}H_{14}$) polyatomic cluster ion implantation doping were reported [22, 23]. These were compared with devices that were fabricated using conventional ion implantation of B^+ and BF_2^+. Using decaborane implantation at 5 kV followed by rapid thermal annealing at 1000°C for 10 s resulted in devices with 39 nm deep junctions. The performance of the devices was equivalent to that of similar devices fabricated using B^+ and BF_2^+ implantation, and the work showed that decaborane offered important advantages for the shallower junction devices to be required in the future. This was the first comparison between cluster implantation and conventional atomic and molecular implantation. The V_{th} roll-off characteristics and I_{off}-I_{on} characteristics of the devices implanted using decaborane were evaluated and showed good drive current ability and excellent short-channel effects. This confirmation of low-energy effect demonstrated the excellent possibilities for future use in ultrashallow junctions and for production fabrication of p-MOSFETs.

Following the initial demonstrated success of doping using $B_{10}H_{14}$, an evaluation was conducted to determine whether the decaborane ion implantation could be employed as an alternative technology for p-MOSFET manufacturing [24]. The conclusion from the evaluation was that decaborane would be practical for production manufacturing of ultrashallow p-type junctions if an ion source able to generate enough beam current could be accomplished. A suitable ion source was then developed, and a high-current ion implanter was modified to incorporate the source. The equipment was used to process 200 mm wafers, and detailed characteristics of fabricated p-MOSFETs with quarter micron ultrashallow junctions were reported [25]. Data from these devices were compared with results from similar devices fabricated using B^+ implantation at 500 eV. The results demonstrated that there were no significant differences in operating characteristics between the devices. This development effort helped push forward the use of decaborane for practical ultrashallow junction formation. Subsequently, several other polyatomic cluster ions such as $B_{18}H_{22}$ [26, 27], $B_{36}H_{44}$ [28], and $C_2B_{10}H_{12}$ were also used for device fabrications [29].

9.4 Analytical Instrumentation Applications

High sputtering yields and low-energy irradiation effects that are inherent properties of GCIB interactions with surfaces

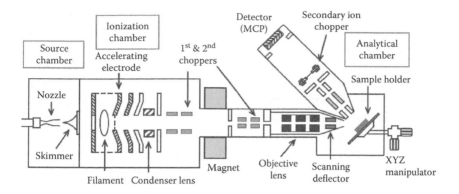

Figure 9.18 Schematic of GCIB TOF-SIMS.

are recognized to be ideal characteristics for applications in surface analysis, especially for complex structures and fragile materials. The combination of minimal ion-mixing behavior and absence of surface-roughening effects is extremely useful for high-resolution depth profiling. Use of a cluster ion source in SIMS apparatus for analysis of surface composition was first investigated by experiment and by molecular dynamics (MD) simulations in 2001 [30]. Using a time-of-flight (TOF) analyzer, a GCIB-SIMS apparatus was reported in 2007 [31, 32]. Figure 9.18 shows a schematic diagram of the equipment. A primary Ar cluster ion beam accelerated at 10–20 kV was size-selected by a chopper using double deflection plates, and was then introduced upon the sample. Secondary ions produced by GCIB were accelerated to 2 kV and were detected with a microchannel plate (MCP). The timing of the secondary ion chopping and the detection were used as the start and stop signals for the TOF measurement.

The first SIMS depth profiling of an organic thin film was conducted with a common amino acid arginine ($C_6H_{14}O_2N_4$) as the target [30]. By choosing cluster acceleration potentials such that energies of constituent Ar atoms of the cluster ions were limited to a few electron volts, fragment ions from the target were rarely observed, and nondissociated molecular ions were readily measured. Subsequent SIMS depth profiling studies involving several materials, such as $(C_8H_8)_n$, $C_{44}H_{88}NO_8P$, $C_{27}H_{18}AlN_3O_3$, and $C_{44}H_{32}N_2$, showed that the intensities of the molecular ions from the samples remained constant with increasing fluence, that is, as the surfaces were being continuously removed. Depth resolution was estimated to be better than 10 nm, and at least two orders of magnitude of dynamic range were obtained [33]. It was concluded that slow Ar cluster ions could be used as powerful tools for SIMS depth profiling analyses of complex molecular materials. It was suggested that the cluster ions could also be applied for precise analyses of organic devices and even biological tissues.

Commercial XPS equipment with a GCIB sputtering source was introduced in 2009 [34]. Due to the recent development of organic devices such as organic molecular transistors, dye-sensitized solar cells, and organic light-emitting devices, material analyses of surface and internal interface chemical structures and functional group structures have become important issues. Conventional Ar ions and C_{60} ions were being used for XPS depth analyses, but irradiations able to be used for profiling with lower damage, less chemical reactions, and without carbon deposition were needed for more precise depth profile analyses. Ar-GCIB was introduced with the expectation that the energy per Ar atom of the GCIB clusters could be suppressed to a level lower than the chemical bond energies of the samples, and unlike Ar ions and C_{60} ions, properly selected GCIB clusters would not cause dissociation, chemical reaction, or carbon deposition within the samples.

Figure 9.19 shows a photograph and a schematic diagram of an Ar-GCIB gun used for the XPS apparatus. The Ar-GCIB gun consists of a cluster generator, an ionizer, extraction electrodes, a Wien filter that uses an electromagnet to remove monomer ions and small cluster ions, a beam deflector that removes fast neutral particles generated by charge exchange, a diaphragm that limits the beam's diameter, a gate valve that isolates the analyzer during maintenance of the ion gun, an Einzel lens for beam convergence, and a scanner for X-Y beam scanning. Results of material analysis of naphthyl phenyl benzene (NPB) films that were used as hole transport layer materials for organic electroluminescence devices have been reported. Ratios of C and N atom compositions before sputtering and after sputtering by GCIB were compared with the composition ratios that resulted with Ar and C_{60} ion beams as XPS sputtering sources. The composition ratios were found to

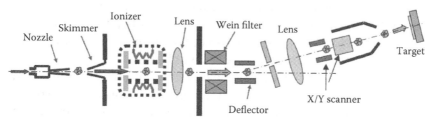

Figure 9.19 Photograph and a schematic diagram of the Ar-GCIB gun for XPS apparatus.

have been significantly changed by sputtering with the C_{60} and Ar ion guns, while they were largely unchanged by sputtering with the Ar-GCIB gun.

9.5 Biomaterial Applications

Rapid progress is being made in the development of many devices that can be implanted into the human body in order to correct a wide range of medical problems. Obvious examples include vascular and neurological stents, dental implants, hip and knee replacement components, spinal fusion cages, etc. Materials that are employed in these devices must always be selected on the basis of their bulk properties being adequate to provide the required functional characteristics needed for the particular device; that is, they must have the proper mechanical strength, elasticity, durability, chemical stability, machinability, and so on. However, upon introduction of a biomedical device into the body, the reaction of the systems of the body to the presence of that new device is largely dependent not upon the bulk characteristics of the materials employed in the device, but rather upon the atomic-level properties of the surfaces of those materials. Undesirable compromises are often necessary in order to accomplish sufficient functional performance by the device while also promoting adequate response to the device by its host. It is recognized that an ability to modify and adequately control the atomic-level characteristics of the surface of a device material, without altering any of the bulk properties of that material, will be great benefit to development of many biomedical devices of the future. Development work now in progress seeks to control and differentiate biological cell actions upon processed surfaces.

In the United States, Exogenesis Corporation was founded in 2005 to utilize GCIB specifically for modification and control of biomedical device surfaces. A number of applications are being developed, and collaborations with biomedical device manufacturer partners are active [35–38]. GCIB is uniquely capable of producing many aggressive modification effects upon any surface without causing any change at all to material more than a few nanometers below the surface. Available actions produced by GCIB include atomic-scale smoothing, roughening and texturing, shallow-layer amorphization, surface chemistry alteration, conversion of hydrophobic character to hydrophilic behavior, and restructuring of molecular bonds and atomic charge states.

In addition to its work to use GCIB for biomedical applications, Exogenesis Corporation has also introduced another concept known as accelerated neutral atom beam (ANAB), in which a modified GCIB configuration is employed to produce intense beams of energetic neutral atoms, with controllable average energies ranging from less than 10 eV to more than 100 eV [39]. GCIB and ANAB have become versatile and complementary techniques for processing of biomaterials, with each technique offering substantial advantages under certain circumstances. In the ANAB technique, a beam of accelerated gas cluster ions is first produced, as is usual in GCIB, but conditions within the GCIB source are altered such that immediately after acceleration, the clusters undergo collisions with nonionized gas atoms. Energy transfer during the collisions causes the clusters to dissociate into individual atoms, and an electrostatic deflector is then used to eliminate the remaining charged species, leaving the released neutral atoms to travel to the process target at the same velocities they had as components of their parent clusters. Upon target impact, the accelerated neutral atom beams produce surface modification effects similar to those normally associated with GCIB, but with appreciably shallower maximum surface penetration depths, typically 2 nm or even less. ANAB is being used for processing of thermally fragile materials such as drugs, for smoothing of surfaces to levels below 0.1 nm Ra, and for introducing nanoscale surface texture into many types of materials. Figure 9.20 shows an Exogenesis nAccel™ system that can be employed for either GCIB or ANAB processing [40]. A basic schematic of the system configuration is shown in Figure 9.21.

Figure 9.20 Exogenesis Corporation nAccel™ equipment for both GCIB and ANAB processing.

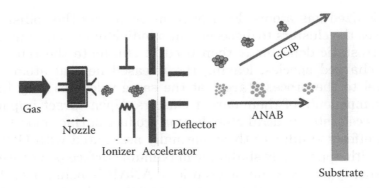

Figure 9.21 Configuration of nAccel™ GCIB and ANAB equipment.

9.5.1 GCIB Processing for Biomaterial Applications

Exogenesis Corporation has described the use of GCIB to control the rate of drug release from drug-coated vascular stents [41]. Conventional practice for producing a drug-eluting stent (DES) has been to combine the drug into a polymer that controls the rate of release, but some DES recipients have been found to experience catastrophic blood clotting or serious inflammatory response issues that are suspected to be associated with influence of the polymer. A method is desired for control of drug elution without using a polymer. The drugs that must be employed are themselves highly soluble in blood and would elute far too quickly in the absence of some method to provide controlled delay of the elution action. Ar-GCIB has been shown to be capable of modifying the outermost surface of a polymer-free drug coating so as to transform a very thin (~10 nm) layer of the drug itself into a carbon-matrix barrier material offering the necessary degree of elution control. The barrier layer is formed because GCIB impact momentarily produces an extreme temperature transient within surface drug molecules, causing those molecules to dissociate, resulting in loss of their gaseous constituents, and leaving behind only the nonvolatile carbon component.

Figure 9.22 shows one simple example demonstration. Bare metal cobalt chrome stents were coated with rapamycin to a thickness of several microns. Some of the stents were processed using Ar-GCIB. All the stents were introduced into

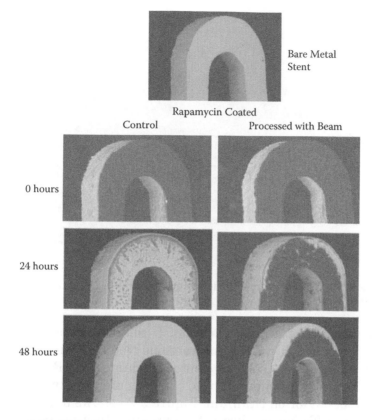

Bare Metal
Stent

Rapamycin Coated

Control

Processed with Beam

0 hours

24 hours

48 hours

Figure 9.22 SEM images showing rapamycin drug coatings on stents without and with Ar-GCIB processing after various times of exposure to human plasma.

human plasma for up to 168 h. As can be seen in the figure, in the case of stents without the GCIB processing, almost all of the rapamycin eluted away within 24 h, and within 48 h none remained, while on the GCIB-processed stents, almost all of the drug remained after 48 h. Subsequently, it was observed that on the GCIB-processed stents, approximately 60% of the drug still remained after 168 h.

GCIB is being employed to modify surfaces of various materials used in biomedical devices with the objective of improving biological responses to the surfaces of the devices, for example, to promote faster healing, more rapid bone growth, stronger integration, and so forth. Teflon is not cytocompatible; that is, it is a material to which cells such as osteoblasts (bone cells)

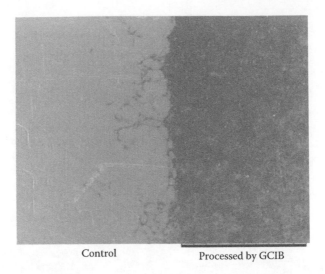

Control   Processed by GCIB

Figure 9.23 Osteoblast cell attachment and proliferation on Teflon surface after 10 days.

normally cannot attach. Osteoblast cells can adhere to titanium, but more rapid cell interactions would be beneficial for many applications, such as dental implants. GCIB is able to make Teflon cytocompatible and can promote greatly enhanced growth behavior on titanium. Figure 9.23 shows a microscope image of a Teflon surface after exposure to a suspension of osteoblast cells for 10 days [39]. Prior to the cell exposure, the right side of the surface had been processed by Ar-GCIB, while the left side was not processed. No cells attached to the unprocessed left side, but a very high density of cells can be seen on the processed side, indicating that the GCIB treatment facilitated strong cell attachment and proliferation.

9.5.2 ANAB Processing for Biomaterial Applications

Polyetheretherketone (PEEK) is a material that is currently gaining popularity in orthopedic and spinal applications. PEEK exhibits good biocompatibility, has elasticity similar to that of human bone, and is radiolucent (transmits x-rays), but for some biomedical applications where it might otherwise be ideal, it has drawbacks because it is inert and does not integrate well with bone. A number of development efforts have been focused upon improving the bioactivity of PEEK by addition of surface coatings, by impregnation of bioactive materials into the PEEK, and by employing various surface modification approaches, such as plasma treatments, plasma immersion ion implantation, and laser irradiation [42–46].

Exogenesis Corporation has described the use of its ANAB technique for surface treatment to enhance the bioactivity of PEEK [47]. ANAB using an inert gas such as Ar leaves no surface residuals following exposure. Upon impact with a surface at low energy and high density, the gases from ANAB dissipate and are immediately pumped away. The result is an extremely pure surface treatment that modifies a very shallow layer on the PEEK surface so as to add bioactive properties. Figure 9.24 shows example AFM images of PEEK surfaces (a) without processing and (b) after Ar-ANAB exposure. The average energy of the ANAB neutral Ar atoms produced from GCIB cluster ions accelerated at 30 kV was approximately 40 eV. The ANAB dose was 2.5×10^{17} atoms/cm^2. A nanoscale texture created by the ANAB exposure is seen in Figure 9.24b. In addition to introducing nanoscale texture, the ANAB treatment also caused the PEEK surface to become strongly hydrophilic. Figure 9.24c and d shows optical images of water droplets on the unprocessed control and ANAB-processed surfaces. The contact angle decreased from 76.4° on the unprocessed surface to 36.1° on the ANAB-treated PEEK. Both surface texture and surface hydrophilicity are considered to be factors important to initial cell attachment action when a device is implanted into the body.

Tests conducted on the ANAB-treated PEEK showed cell attachment and proliferation characteristics to be comparable to those that occur on untreated titanium, which is considered to be a good cytocompatible material [47]. Figure 9.25 shows osteoblast cell counts on titanium and unprocessed PEEK control surfaces and on ANAB-processed PEEK surfaces at 3, 7, and 10 days following exposure to osteoblast cell suspensions.

As proof of concept that ANAB surface modification leads to better living body integration, an *in vivo* study was performed in which unprocessed and ANAB-processed PEEK disks were implanted into rat skulls. Four weeks later, histology was performed to examine the amount of bone regrowth onto the PEEK surfaces. It was found that in the case of the ANAB-treated

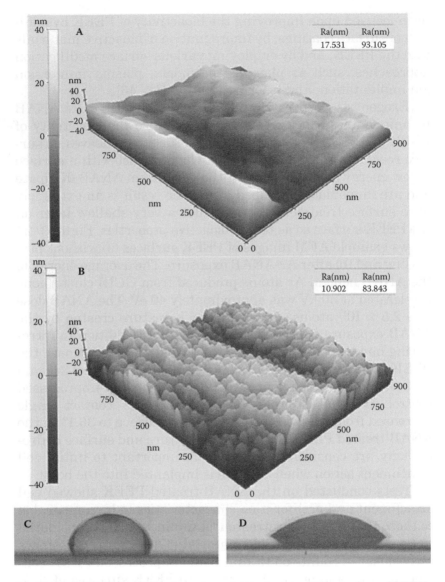

Figure 9.24 ANAB results in nanoscale texturing of the surface of PEEK. AFM measurements revealed the generally (A) smooth surface of PEEK results in (B) nanotexturing following ANAB treatment. ANAB also results in increased PEEK surface hydrophilicity as measured by water contact angle. (C) Control PEEK surface has a contact angle of $\theta = 76.4°$, while (D) ANAB-treated PEEK has a contact angle of $\theta = 36.1°$.

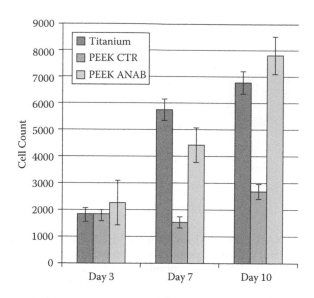

Figure 9.25 ANAB treatment of PEEK results in cell proliferation comparable to that of native Ti. Human fetal osteoblast (hFOB) cells growing on surfaces of ANAB-treated PEEK display rapid proliferation compared to control surfaces. ANAB results in increasing the bioactivity of PEEK, comparable cell attachment, and proliferation is seen on bioactive Ti. There were no significant differences in cell proliferation between the Ti and ANAB-treated PEEK groups at any time point.

PEEK, bone ledge had grown to cover approximately 50% of the PEEK surface, whereas on the unprocessed control PEEK, no bone had formed. Histology images from this rat calvarial critical defect model study are shown in Figure 9.26 [47].

Exogenesis has reported another animal study to further characterize the ability of ANAB surface modification to enhance the bioactivity and osseointegration ability of PEEK [48]. Control and ANAB-treated PEEK plugs were implanted into the cortical region of the femur and the cancellous regions of the femur and tibia in sheep for periods of 4 and 12 weeks. Some plugs included 1 mm³ surface indent pits introduced for the purpose of allowing observation of the ability of new bone to fill in a void. At the appropriate time points, the sheep were sacrificed and the PEEK implants with surrounding bone were assayed by microcomputed tomography (μCT), and they were

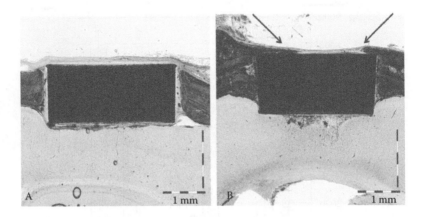

Figure 9.26 ANAB enhances bioactivity of PEEK and results in bone integration *in vivo*. Rat calvarial defect study reveals control PEEK has very little ability to integrate with surrounding bone. (A) Only a fibrous layer is seen by histopathological methods on control PEEK disks, and further bone resorption is evident on the edges where bone makes initial contact with PEEK. (B) ANAB-treated PEEK disks result in very good purchase on the contact site between bone and PEEK, as well as a cortical bone ledge that appears to cover nearly 50% of the surface by week 4 (indicated by arrows).

further processed for histology. At the 4-week time point, it was evident that the unprocessed control PEEK implants were already displaying signs of device failure. Fibrous tissue had begun to envelop the unprocessed PEEK implant. It is known that such fibrous tissue results in weak attachment between the bone and the surface of the implant, and ultimately will lead to interface failure. The ANAB-treated implants showed very good bone adhesion to the surface, complete absence of fibrous tissue, and the start of growth progression into the indent pit voids. As can be seen by inspection of Figures 9.27 and 9.28, at the end of 12 weeks, both histology and μCT confirmed that the ANAB-treated PEEK implants resulted in complete void fill, while the unprocessed controls lacked any bone growth in the voids. The ANAB treatment unquestionably increased the ability of the PEEK to osseointegrate.

The studies performed at Exogenesis showed that GCIB or ANAB processing to enhance bioactivity can make it feasible to use alternate materials with more favorable characteristics for implantable medical devices where such materials would otherwise not be usable because they would fail to integrate.

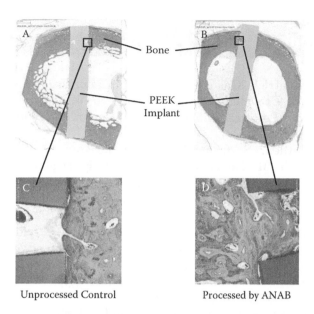

Figure 9.27 Stained histology images of PEEK implants after 12 weeks: (A, D) unprocessed control showing absence of bone growth into indent void and (B, D) treated by ANAB exhibiting complete void fill.

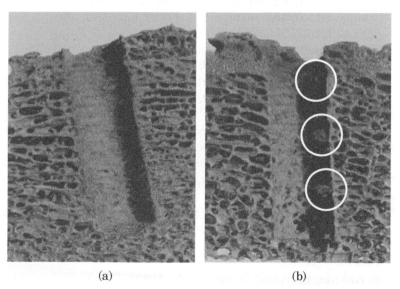

Figure 9.28 μCT reconstruction of radiolucent PEEK and surrounding bone: (a) control and (b) ANAB-processed PEEK at 12 weeks postimplantation. Bone growing into the gaps is seen on the ANAB-treated implants as indicated by the white circles (b), which is absent with controls (a).

References

1. A. Kirkpatrick. Gas cluster ion beam applications and equipment. In *Extended Abstracts of the 3rd Workshop on Cluster Ion Beam Process Technology*, Kyoto, Japan, September 10–12, 2002, pp. 21–31.
2. L. P. Allen, S. Caliendo, N. Hofmeester, E. Harrington, M. Walsh, M. Tabat, T. G. Tetreault, E. Degenkolb, and C. Santeufemio. SOI uniformity and surface smoothness improvement using GCIB processing. In *IEEE International SOI Conference*, Williamsburg, Virginia, 2002, pp. 192–193.
3. C. Eggs, E. Schmidhammer, and A. Schäufele. Yield enhancement for BAW production using local corrective etching. In *7th Workshop on Cluster Ion Beam Process Technology and Quantum Beam Process Technology*, Tokyo, Japan, November 6–7, 2006, pp. 46–51.
4. T. Mashita, N. Toyoda, and I. Yamada. Surface smoothing of polycrystalline substrates with gas cluster ion beams. *Jpn. J. Appl. Phys.*, 49, 06GH09-1–06GH09-3, 2010.
5. E. Bourelle, A. Suzuki, A. Sato, T. Seki, and J. Matsuo. Polishing of sidewall surfaces using a gas cluster ion beam. *Jpn. J. Appl. Phys.*, 43, L1253–L1255, 2004.
6. NEDO. Advanced nano-fabrication process technology using quantum beams (in Japanese). NEDO Project Report P02048. New Energy and Industrial Technology Development Organization, October 31, 2007.
7. I. Yamada, J. Matsuo, and N. Toyoda. Summary of industry-academia collaboration projects on cluster ion beam process technology. In *Ion Implantation Technology: 17th International Conference on Ion Implantation Technology*, vol. 1066, 415–418. AIP Conference Proceedings, American Institute of Physics, New York, 2008.
8. NEDO. Advanced nano-fabrication process technology using quantum beams (in Japanese). Final evaluation report. New Energy and Industrial Technology Development Organization, February 2008.
9. S. Kakuta, S. Sasaki, T. Hirano, K. Ueda, T. Seki, S. Ninomiya, and M. Hada. Low damage smoothing of magnetic material films using a gas cluster ion beam. *Nucl. Instrum. Methods Phys. Res. B*, 257, 677–682, 2007. Figure of HDD sensor structure was shown on the poster of the SMMIB 2005.

10. N. Toyoda and I. Yamada. Fabrication of patterned media by using gas cluster ion beams. In *10th Workshop on Cluster Ion Beam Process Technology and Quantum Beam Process Technology*, Kyoto, Japan, June 14–15, 2010, pp.75–78.

11. T. Seki, Y. Yoshino, T. Senoo, K. Koike, S. Ninomiya, T. Aoki, and J. Matsuo. High speed Si etching with ClF_3 cluster injection. In *18th International Conference on Ion Implantation Technology IIT 2010*, eds. J. Matsuo, M. Kase, T. Aoki, and T. Seki, 317–324. AIP Conference Proceedings 1321. American Institute of Physics, New York, 2010.

12. K. Koike, Y. Yoshino, T. Senoo, T. Seki, S. Ninomiya, T. Aoki, and J. Matsuo. Anisotropic etching using reactive cluster beams. *Appl. Phys. Exp.*, 3(12), 126501-1–126501-3, 2010.

13. N. Variam, S. Falk, and S. Mehta. Challenges and solutions in the process integration of ultra-shallow junctions in advanced CMOS technology. In *Proceedings of the 13th International Conference on Ion Implantation Technology*, 77–80. Institute of Electrical and Electronics Engineers, Piscataway, NJ, 2000.

14. H.-J. L. Gossmann. Ion implantation in advanced planer and vertical devices. *Nucl. Instrum. Methods Phys. Res. B*, 237, 1–5, 2005.

15. H. Iwai. Roadmap for 22 nm and beyond. *Microelectr. Eng.*, 86, 1520–1528, 2009.

16. M. Koyanagi. Requirement for junction technology from device design. In *Extended Abstracts of the First International Workshop on Junction Technology*, Makuhari, Chiba, Japan, 2000, pp.1-1-1–1-1-6.

17. J. M. Poate, A. Agarwal, and L. M. Rubin. Challenges for ion implantation. In *Extended Abstracts of the Second International Workshop on Junction Technology (IWJT)*, Tokyo, Japan, 2001, pp. 1-1-1–1-1-5.

18. J. Hautala, J. Borland, M. Tabat, and W. Skinner. Infusion doping for USJ formation. In Fourth International Workshop on Junction Technology (IWJT '04), 50–53. Institute of Electrical and Electronics Engineers, Piscataway, NJ, 2004.

19. W. Skinner, M. Gwinn, J. Hautala, and T. Kuroi. Infusion processing for advanced transistor manufacturing. In *2006 IEE/SEMI Advanced Semiconductor Manufacturing Conference*, Boston, May 22–24, 2001, pp. 214–218.

20. T. Yamashita, T. Hayashi, Y. Nishida, Y. Kawasaki, T. Kuroi, H. Oda, T. Eimori, and Y. Ohji. Formation of S/D-extension using boron gas cluster ion beam doping for sub-50-nm PMOSFET.

In *Fifth International Workshop on Junction Technology (IWJT 2005)*, 35–36. Institute of Electrical and Electronics Engineers, Piscataway, NJ, 2005.

21. M. Kitazawa, Y. Kawasaki, M. Togawa, D. Rosser, T. Yamashita, T. Iwamatsu, H. Oda, and Y. Inoue. Sub-30-nm PMOSFET using gas cluster ion beam boron doping for 45-nm node CMOS and beyond. In *Seventh International Workshop on Junction Technology (IWJT '07)*, 61–62. Institute of Electrical and Electronics Engineers, Piscataway, NJ, 2007.

22. D. Takeuchi, N. Shimada, J. Matsuo, and I. Yamada. Shallow junction formation by polyatomic cluster ion implantation. In *Proceedings of the 11th International Conference on Ion Implantation Technology—IIT '96*, vol. 1, issue 1, pp. 772–775. Institute of Electrical and Electronics Engineers, Piscataway, NJ, 1996.

23. K. Gotoh. A study of sub-0.1µm CMOS process technology, (in Japanese). PhD thesis, Tohoku University, 1998.

24. D. C. Jacobson, K. Bourdelle, H.-J. Gossmann, M. Sosnowski, M. A. Albano, V. Babaram, J. M. Poate, A. Agarwal, A. Perel, and T. Horsky. Decaborane, an alternative approach to ultra low energy ion implantation. In *Proceedings of the 13th International Conference on Ion Implantation Technology*, 300–303. Institute of Electrical and Electronics Engineers, Piscataway, NJ, 2000.

25. A. S. Perel, W. Krull, and D. Hoglund. Decaborane ion implantation. In *Proceedings of the 13th International Conference on Ion Implantation Technology*, 304–307. Institute of Electrical and Electronics Engineers, Piscataway, NJ, 2000.

26. Y. Kawasaki, T. Kuori, T. Yamashita, K. Horita, T. Hayashi, M. Ishibashi, M. Togawa, Y. Ohno, M. Yoneda, T. Horsky, D. Jaconson, and W. Krull. Ultra-shallow junction formation by $B_{18}H_{22}$ ion implantation. *Nucl. Instrum. Methods Phys. Res. B*, 237, 25–29, 2005.

27. Y. Kawasaki. Application of boron cluster ion implantation to PMOSFET. PhD thesis, Hiroshima University, 2012.

28. K. Sekar, W. Krull, K Huet, C. Boniface, and J. Venturini. Large cluster boron ($H_{36}H_x$) implant for USJ applications. In *Proceedings of the 18th International Conference on Ion Implantation Technology*, vol. 1321, 101–104. AIP Conference Proceedings. American Institute of Physics, New York, 2010.

29. A. Renau. A better approach to molecular implantation. In *Seventh International Workshop on Junction Technology (IWJT '07)*, 107–112. Institute of Electrical and Electronics Engineers, Piscataway, NJ, 2007.

30. N. Toyoda, J. Matsuo, T. Aoki, S. Chiba, I. Yamada, D. B. Fenner, and R. Tori. Secondary ion mass spectrometry with gas cluster ion beams. *Mater. Res. Soc. Proc.*, 647, O5.1.1–O5.1.6, 2001.
31. S. Ninomiya, K. Ichiki, Y. Nakata, Y. Honda, T. Seki, T. Aoki, and J. Matsuo. Ionization and low damage etching of soft materials with slow Ar cluster ions. In *8th Workshop on Cluster Ion Beam Process Technology and Quantum Beam Process Technology*, Tokyo, Japan, November 8–9, 2007, pp. 41–46.
32. S. Ninomiya, Y. Nakata, K. Ichiki, T. Seki, T. Aoki, and J. Matsuo. Measurement of secondary ions emitted from organic compounds bombarded with large gas cluster ions. *Nucl. Instrum. Methods Phys. Res. B*, 256, 493–496, 2007.
33. S. Ninomiya, K. Ichik, T. Seki, T. Aoki, and J. Matsuo. Characteristics of organic depth profiling using large cluster ion beams. In *10th Workshop on Cluster Ion Beam Process Technology and Quantum Beam Process Technology*, Kyoto, Japan, June 14–15, 2010, pp. 43–47.
34. T. Kunibe, D. Sakai, K. Mamiya, T. Yuze, N. Sanada, and S. Shimizu. A sputter gun that uses gas cluster ion beams for a low damage depth profile analysis. In *9th Workshop on Cluster Ion Beam Technology*, Tokyo, Japan, March 11–12, 2009, pp. 57–60.
35. J. Khoury, G. C. Kodali, L. Tarrant, R. C. Svrluga, and S. R. Kirkpatrick. Surface modification by gas cluster ion beam (GCIB) as a novel drug delivery method for vascular stents. *FASEB J.*, April 6, 2010, abstract 644.9.
36. J. Khoury, S. R. Kirkpatrick, G. C. Kodali, R. C. Svrluga, and L.-J. B. Tarrant. Gas cluster ion beam (GCIB) surface modification of titanium enhances osteoblast proliferation and bone formation *in vitro. FASEB J.*, April 6, 2010, abstract 638.4.
37. I. Yamada and J. Khoury. Cluster ion beam processing: Review of current and prospective applications. *Mater. Res. Soc. Symp. Proc.*, 1354, 21–31, 2011.
38. http://www.exogenesis.us/platform-technology.
39. A. Kirkpatrick, S. Kirkpatrick, M. Walsh, S. Chau, M. Mack, S. Harrison, R. Svrluga, and J. Khoury. Investigation of accelerated neutral atom beams created from gas cluster ion beams. *Nucl. Instrum. Methods Phys. Res. B*, 307, 281–289, 2013.
40. http://www.exogenesis.us/sales-and-services.
41. Exogenesis. Controlled elution of drugs and other therapeutic compounds. White paper. http://www.exogenesis.us/wp-content/uploads/2013/05/Drug-Elution-without-Polymers-White-Paper.pdf.

42. S. M. Kurtz and J. N. Devine. PEEK biomaterials in trauma, orthopedic, and spinal implants. *Biomaterials*, 28, 4845–4869, 2007.
43. M. M. Bilek and D. R. McKenzie. Plasma modified surfaces for covalent immobilization of functional biomolecules in the absence of chemical linkers: Towards better biosensors and a new generation of medical implants. *Biophys Rev.*, 2, 55–65, 2010.
44. S. M. Kurtz (ed.). *PEEK Biomaterials Handbook*. William Andrew, Kidlington, Oxford, UK, 2012.
45. R. Ma and T. Tang. Current strategies to improve the bioactivity of PEEK. *Int. J. Mol. Sci.*, 15, 5426–5445, 2014.
46. J. Waser-Althaus. Plasma-activated polymer films for mesenchymal stem cell differentiation. PhD thesis, University Basel, 2014.
47. J. Khoury, S. R. Kirkpatrick, M. Maxwell, R. E. Cherian, A. Kirkpatrick, and R. C. Svrluga. Neutral atom beam technique enhances bioactivity of PEEK. *Nucl. Instrum. Methods Phys. Res. B*, 307, 630–634, 2013.
48. Courtesy of Dr. Joe Khoury, Exogenesis Corporation.

10
Conclusions

During the decades that have passed since gas cluster ion beam (GCIB) technology first began in 1988, great progress has been made in the development of the equipment and processes. Various types of equipment have been developed for smoothing, etching, trimming, shallow implantation, thin-film deposition, surface analysis applications, and other surface modification or surface control purposes. Techniques for sub-nanometer-scale processing of metals, semiconductors, and insulating materials have been established and are now being increasingly employed in production.

In the book, the significant advances and technical details have been described. An overview of the history of cluster beam technology has been presented for the purpose of illustrating the differences, and often the uniqueness, of cluster ion beam technology relative to other ion beam techniques. This book illustrates how early ion beam discoveries led to the development of traditional ion implantation technology during the second half of the twentieth century, and explains how early discoveries related to supersonic nozzles and gas condensation led to the creation of gas cluster beams. The knowledge of how cluster ion beam technology has been developed will provide valuable contributions to support further advances in the future.

The presentation of the history and milestones of the cluster beam technology has explained how it has been established as a result of many important historical events, which at first glance would not seem to have been related to the cluster beam technology, but in fact were essential steps toward it.

One example was the recognition of nucleation and condensation phenomena that occurred in steam turbine nozzles so as to interfere with the operation of the turbines, but which later became the basis for cluster generation. The earliest intended technological application for a cluster beam was as an injector for a fusion system. Although the application itself could not be realized, many advances related to the scientific fundamentals of cluster beam formation were made. Similarly, a metal vapor cluster ion beam deposition technique that was a predecessor to GCIB was not successful, but it had much influence toward bringing cluster ion technology to readiness for industrial applications. Additional examples have included the results of other investigations related to supersonic beam formation, the development of various sophisticated techniques for the evaluation of ion beams and beam interaction characteristics, and many advances made in the technology of ion implantation for semiconductors. From this foundation, cluster beam technology investigations were started, and over time, the technology as it exists today has been realized.

To establish the technology, many fundamental cluster ion–solid interaction phenomena were studied experimentally and by molecular dynamics simulations. The results of these studies provided confidence that the technology could be employed in processes of industrial importance. Capable machines and processes were established, and adoption of the technology into industry occurred. To make this happen successfully required more than only the development that took place within individual universities and companies, it also required large-scale carefully planned and coordinated academia-industry collaborations. These were generously supported by the government of Japan.

Most recently, the GCIB concept has continued to be adapted into new technological variations that can be used in subnanometer etching and smoothing and for new surface modification applications in metals, semiconductors, insulators, and organic materials. Extensive R&D efforts have been initiated toward practical industrial uses in surface analytical instrumentation applications such as x-ray photoelectron spectroscopy (XPS) and secondary ion mass spectroscopy (SIMS) and for surface modification of biomaterials. However, while the technology is now expanding into new areas such as physical and chemical etching of organic and inorganic materials, the fundamental features of some of the new approaches have not yet been explored, and the concepts involved still remain relatively new and very different from those associated with well-known traditional ion beam techniques. The author expects that future research and development by persons who become interested in this technology, and the new opportunities that it can offer, will open a wide range of valuable new material processing applications.

Index

H

I

J

T

V

W

X

Y

U

Z

Printed and bound by CPI Group (UK) Ltd, Croydon, CR0 4YY

01/11/2024

01782614-0005